The ERGONOMIC CASEBOOK

Real World Solutions

James P. Kohn

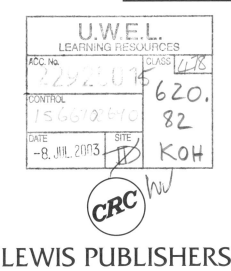
LEWIS PUBLISHERS

Boca Raton   New York   London   Tokyo

Acquiring Editor: Ken McCombs
Project Editor: Albert W. Starkweather, Jr.
Cover Designer: Dawn Boyd
Prepress: Greg Cuciak
Marketing Manager: Greg Daurelle
Manufacturing Assistant: Sheri Schwartz

Library of Congress Cataloging-in-Publication Data

The ergonomic casebook: real world solutions / James P. Kohn.
    p.   cm.
    Includes bibliographical references and index.
    ISBN 1-56670-269-0 (alk. paper)
    1. Human engineering — case studies.   I. Title.
TA166.K47   1997
620.8'2—dc21   96-37140
                                                    CIP

© 1997 by CRC Press, Inc.
Lewis Publishers is an imprint of CRC Press

No claim to original U.S. Government works
International Standard Book Number 1-56670-269-0
Library of Congress Card Number 96-37140
Printed in the United States of America 1 2 3 4 5 6 7 8 9 0
Printed on acid-free paper

About the Author

James P. Kohn is an associate professor of Industrial Technology at East Carolina University. He was previously director of the Occupational Safety Management program at Indiana State University. Dr. Kohn holds a doctorate in education from West Virginia University. He has both academic and industrial experience. He has worked as a safety, health and environmental consultant in the telecommunications for industry and as a corporate supervisor of safety and health training in the electrical utility industry. In the area of occupational safety and health he is a Certified Safety Professional (CSP), Certified Industrial Hygienist (CIH), and Certified Professional Ergonomist (CPE).

Dr. Kohn has written four books on safety and industrial hygiene topics. He has also published numerous articles covering ergonomics, safety, industrial hygiene and training. In addition, Dr. Kohn has presented numerous papers at professional conferences and has conducted many training workshops both industry and the public sector.

He is a member of the American Society of Safety Engineers, American Industrial Hygiene Association, and the American Conference of Governmental Industrial Hygienists. He currently serves as a member of the Eastern Carolina Safety and Health School Board.

Acknowledgments

The author wishes to thank all of his professional colleagues and students who contributed to the development and editing of this manuscript. This project would not have been possible were it not for the contributions made by the authors of the case studies. Special thanks go to Mark Friend and Celeste Winterberger for their contribution to this project as well as their encouragement and support. Special thanks also go to Jennifer Lewis and Leslie Ridings for editing this manuscript. In addition, thanks goes to Mr. Randy Godwin for assisting in the computer scanning of the photographs included in this work.

I would also like to thank my wife, Carrie Kohn M.A., for her assistance, patience, and support during the development of this manuscript.

Preface

One of the greatest educational challenges in the Occupational Health And Safety Profession is the application of theory to actual workplace practice. It is one thing to discuss how workstations should be ideally designed and quite another thing to address the real world limitations associated with pre-existing facilities. I have attempted to bridge this gap in the ergonomic courses that I conduct by employing the use of case studies.

This very popular activity of using case studies in the classroom augments the theory presented in lectures to real world problems that companies face on a daily basis. In these case studies methods are described for identifying ergonomic problems. Specific causes, typically referred to as ergonomic stressors, are then reported. Recommended strategies for the elimination of identified stressors are indicated. Implemented strategies and evaluated results are frequently discussed when available. Students and professionals using these case studies reported that this activity helped them to bridge the academic gap and facilitate comprehension of ergonomic application principles.

The purpose of this publication is to share with individuals new to the ergonomics discipline examples of ergonomic anticipation, recognition, evaluation, and control concepts. Instructors may wish to use this resource as a supplement to their traditional ergonomic textbook. They may use the case studies found in this publication as a tool for explaining how ergonomic principles are applied to the occupational environment. Instructors may also wish to have students develop their own case studies based upon the published examples.

This resource is not restricted to use in the classroom. This publication will also be useful to the health and safety professional with limited expertise in ergonomics. Those individuals can review the following case studies for ideas to assist them in solving ergonomic problems that they face in their organization. The possible application of this resource is endless. Ultimately, both student and professional should gain extensive insight into ergonomic problem-solving as a result of the case studies found in this publication.

James P. Kohn, EdD, CSP, CIH, CPE
Greenville, NC

Table Of Contents

Chapter 1. Introduction, J. Kohn and C. Winterberger **1-1**

    What Is Ergonomics?   1-1

        Terminology..   1-2

    Why Is Ergonomics Important?   1-2

    How Big a Problem is Ergonomics?   1-4

        Medical Costs..   1-5

        Legal Costs   1-6

        Costs to Organizations   1-6

        Benefits   1-7

        Ergonomic areas of importance   1-8

        Psychological   1-9

        Behavioral   1-9

        Psychosocial   1-10

    What Are Some Strategies for Identifying Ergonomic

        Problems?   1-10

    How Does Human Physiology Affect Ergonomics?   1-11

        Micro Physiological Response   1-12

        Macro Physiological Concerns   1-16

    What Are Some Ergonomically Related Injuries?.......   1-17

        Cumulative Trauma Disorders   1-17

        Musculoskeletal Disorders   1-19

        Physiological Stress Disorders   1-25

        Causes of Cumulative Trauma Disorders   1-26

    Are Some Industries More Likely To Have

        Ergonomic Injuries?   1-27

        Meat Packing   1-27

        Office Environment   1-28

        Construction   1-29

        Manufacturing/assembly   1-31

    What Are Some Methods To Eliminate Or Control

        Ergonomic Problems?   1-33

        Human Control Methods   1-33

        Human Environmental Control Methods   1-34

        Equipment/Facility Control Methods   1-35

        Organizational Control Methods   1-36

    What Can Be Learned From The Experience Of

        Others?   1-36

    Why Case Studies?   1-37

    How Is The Case Study Book Organized?   1-37

    Conclusion   1-38

    References   1-39

Chapter Questions 1-43
Chapter 2.1 *Agriculture Case Studies* **2.1-1**
  Ergonomic Job Design in the Hog Industry,
    W. Hilton **2.1-2**
    Introduction 2.1-2
    Case Study 2.1-2
    Recommendations 2.1-6
    References... 2.1-6
    Case Study Questions 2.1-7
  Overexertion in the Hog Industry, W. Hilton **2.1-8**
    Introduction 2.1-8
    Case Study 2.1-8
    Recommendations 2.1-11
    References... 2.1-11
    Case Study Questions 2.1-13
Chapter 2.2 *Chemical Industry Case Studies* **2.2-1**
  Anthropometrics in the Chemical Industry,
    J. Kohn **2.2-2**
    Introduction 2.2-2
    Statement of the Problem 2.2-2
    Case Study 2.2-2
    Recommendations 2.2-4
    References... 2.2-5
    Case Study Questions 2.2-5
  Hand Injuries in a Chemical Industry Quality
    Control Laboratory, J. Kohn **2.2-6**
    Introduction 2.2-6
    Statement of the Problem 2.2-6
    Case Study 2.2-6
    Recommendations 2.2-8
    References... 2.2-8
    Case Study Questions 2.2-9
Chapter 2.3 *Electronics Industry Case Studies* **2.3-1**
  Push, Pull, and Lifting Material Handling Issues
    in the Electronics Industry, J. Kohn **2.3-2**
    Introduction 2.3-2
    Statement of the Problem 2.3-2
    Case Study 2.3-2
    Recommendations 2.3-3
    References... 2.3-4
    Case Study Questions 2.3-5
  Repetitive Motion Problems In A Small Electric
    Motor Manufacturing Facility, M. Friend **2.3-6**
    Introduction 2.3-6

Statement of the Problem 2.3-6

Case Study 2.3-6

Recommendations 2.3-7

Case Study Questions 2.3-9

Chapter 2.4 *Hospital/Health Care Case Studies* **2.4-1**

Elimination of Back Injuries in a Hospital
Environment, A. Anderson, B. Maxwell, and
E. Johnson **2.4-2**

Introduction 2.4-2

Statement of the Problem 2.4-2

Case Study 2.4-3

Recommendations 2.4-4

References... 2.4-5

Case Study Questions 2.4-5

An Analysis of Signal Detection Issues in a
Hospital Environment, G. Baker and R. Scott .. **2.4-6**

Introduction 2.4-6

Case Study 2.4-6

Recommendations 2.4-8

References... 2.4-8

Case Study Questions 2.4-9

Manual Lifting Case Studies in the Health Care
Industry, B. Maxwell and E. Johnson **2.4-10**

Introduction 2.4-10

Statement of the Problem No. 1 .... 2.4-10

Recommendations for Problem No. 1 .... 2.4-10

Statement of the Problem No. 2 .... 2.4-11

Recommendations for Problem No.2 .... 2.4-11

Case Study Questions 2.4-12

Psychosocial Case Study of a Medical
Transcription Unit, K. Miezio **2.4-13**

Introduction 2.4-13

Statement of the Problem 2.4-14

Case Study 2.4-14

Recommendations 2.4-16

Case Study Questions 2.4-18

Chapter 2.5 *Manufacturing Case Studies* **2.5-1**

A Behavioral Engineering Approach to Ergonomic
Problems, J. Kohn **2.5-2**

Introduction 2.5-2

Method 2.5-6

Results 2.5-9

Recommendations 2.5-10

References... 2.5-13

Case Study Questions 2.5-15
Ergonomic Case Study of an Automobile Radiator
    Parts Manufacturing Facility, D. Hinnant **2.5-17**
Introduction 2.5-17
Case Study 2.5-17
Recommendations 2.5-19
Case Study Questions 2.5-20
Repetitive Motion Problems in a Synthetic Fibers
    Plant, D. Lawhorn **2.5-21**
Introduction 2.5-21
Case Study 2.5-24
Recommendations 2.5-25
Conclusion 2.5-26
References... 2.5-28
Case Study Questions 2.5-29
Repetitive Motion Problems in Handpacking
    Systems, L. Ridings **2.5-30**
Introduction 2.5-30
Case Study 2.5-31
Results 2.5-32
Recommendations 2.5-32
References... 2.5-33
Case Study Questions 2.5-34
An Ergonomic Case Study in a Servo-Motor
    Assembly Facility, J. Coltrain and B. Dail **2.5-35**
Introduction 2.5-35
Statement of the Problem 2.5-35
Case Study 2.5-35
Recommendations 2.5-36
References... 2.5-37
Case Study Questions 2.5-37
Ergonomic Problems in the Product Design
    Industry, B. Eudy **2.5-38**
Introduction :... 2.5-38
Statement of the Problem 2.5-38
Case Study 2.5-38
Recommendations 2.5-39
Conclusion 2.5-40
Case Study Questions 2.5-40
Chapter 2.6 *Mercantile Case Studies* **2.6-1**
Ergonomic Force Requirements to Open Doors,
    M. Volney **2.61-2**
Introduction 2.61-2
Statement of the Problem 2.61-2

Case Study 2.61-2
Evaluation 2.61-5
Recommendations 2.61-6
References.. 2.61-7
Case Study Questions 2.61-8

Chapter 2.7 *Office Environments Case Studies* **2.7-1**

Ergonomic Complaints Regarding Video Display
    Terminals, L. Ridings **2.7-2**
Introduction 2.7-2
Case Study 2.7-3
Recommendations 2.7-4
Conclusion.. 2.7-5
References.. 2.7-6
Case Study Questions 2.7-7

An Ergonomic Program for Solving VDT
    Workstation Problems, S. Joyner **2.7-8**
Introduction 2.7-8
Case Study 2.7-8
Conclusion 2.7-9
Case Study Questions 2.7-10

An Ergonomic Systems Approach for Solving VDT
    Workstation Problems in the Agricultural
    Industry, J. Kohn **2.7-11**
Introduction 2.7-11
Case Study 2.7-11
Recommendations 2.7-13
Conclusion.. 2.7-13
Case Study Questions 2.7-15

Office Ergonomics and Engineering Design,
    M. Hansen **2.7-16**
Introduction 2.7-16
Case Study 2.7-25
Recommendations 2.7-28
Conclusions 2.7-33
References.. 2.7-34
Case Study Questions 2.7-35

Chapter 2.8 *Service Industry Case Studies* **2.8-1**

Manual Materials Handling in the Frozen Food
    Delivery Industry, N. Brown **2.8-2**
Introduction 2.8-2
Statement of the Problem 2.8-3
Results 2.8-3
Recommendations 2.8-4
Conclusion 2.8-7

References.. 2.8-7
Case Study Questions 2.8-8
Manual Materials Handling: Loading of a
  Freezer in the Frozen Food Delivery Industry,
  N. Brown **2.8-9**
Introduction 2.8-9
Statement of the Problem 2.8-9
Results 2.8-10
Conclusions 2.8-12
Recommendations 2.8-13
References.. 2.8-15
Case Study Questions 2.8-17
An Environmental-Ergonomic Case Study at a
  Parcel Delivery Facility, A. Tillett **2.8-18**
Introduction 2.8-18
Case Study 2.8-18
Recommendations 2.8-19
Case Study Questions 2.8-21
Ergonomic Tool Design Hazards in the Restaurant
  Industry, J. Krause **2.8-22**
Introduction 2.8-22
Statement of the Problem 2.8-22
Case Study 2.8-23
Recommendations 2.8-24
Case Study Questions 2.8-27
Workplace Equipment and Facility Redesign in
  the Restaurant Industry, J. Krause and R. S.
  Lawson **2.8-28**
Introduction 2.8-28
Statement of the Problem 2.8-28
Case Study 2.8-29
Recommendations 2.8-30
Case Study Questions 2.8-33
Chapter 2.9 *Utility Industry Case Studies* **2.9-1**
Signal Detection — Three Mile Island Accident:
  Right String Wrong Yo Yo, R. Pate **2.9-2**
Introduction 2.9-2
Statement of the Problem 2.9-3
References.. 2.9-5
Case Study Questions 2.9-5
Index **3-1**

List of Figures and Tables

Table 1-1:   The 15 private industries with the highest
                 repeated trauma incident rates with SIC Codes
                 for 1994 1-5
Table 1-2 :   The benefits of implementing ergonomics
                 programs.. 1-8
Table 1-3:   The basic physiological systems typically affected
                 by ergonomic stressors 1-12
Table 1-4:   Several structures found in the cell 1-13
Table 1-5:   The lobes of the brain where electrical impulses
                 are sent 1-14
Table 1-6:   The six types of body movements which can
                 occur around joints 1-16
Table 1-7:   Industries having the highest incident rates for
                 common injuries per 10,000 workers in 1994
                 from the Bureau of Labor Statistics 1-19
Table 1-8:   Typical symptoms of cumulative trauma 1-22
Table 1-9:   Ergonomic repetitive motion problems identified
                 in the OSHA Ergonomic Program Management
                 Guidelines for Meatpacking Plants 1-27
Figure 2.1-1.   An employee, about the same height as the
                 injured employee mentioned in the case study,
                 attempts to grab and operate the lever 2.1-4
Figure 2.1-2.   Close up view of the incorrectly installed feed
                 lever system 2.1-4
Figure 2.1-3.   A view of a correctly installed feed lever 2.1-5
Table 2.1-1.   The percentage of serious injuries caused by
                 lifting, pushing, and (overexertion) in the seven
                 major occupational areas in agriculture 2.1-9
Figure 2.1-4.   A picture of the cart-boy, a device used to lift and
                 transport dead hogs/sows out of the finishing
                 house 2.1-12
Table 2.5-1.   Ergonomic hazards by task in the Clear Lacquer
                 Paint and Bundling Areas 2.5-11
Figure 2.5-1.   Diagram of the equipment found in the tube
                 preparation area 2.5-26
Figure 2.5-2:   The top photograph shows the stacks of tubes as
                 delivered to the tube printing area. The bottom
                 photograph shows an employee grasping several
                 tubes at a time to load the tube printing machine 2.5-27
Figure 2.6-1.   A view of the two glass entry doors from outside
                 the mini-mall 2.6-3

Figure 2.6-2.   A view of the two glass entry doors from inside
                 the mini-mall 2.6-3

Figure 2.6-3.   A close up view of the closer jamb located on the
                 top left corner of the glass door 2.6-4

Table 2.6-1.   Arm strength and push/pull capabilities at
                 selected degrees of flexion and selected age
                 ranges 2.6-5

Table 2.7-1.   Repetitive Motion Injuries, Number of Cases 2.7-18

Table 2.7-2:   Measures taken to Address Workplace
                 Ergonomics. 2.7-22

Figure 2.7-1.   Example Ergonomics Checklist for a Chair for a
                 Seated Environment 2.7-29

Figure 2.7-2:   Top View of the Architect's Initial Design 2.7-30

Figure 2.7-3:   Side view of the architect's initial design 2.7-31

Figure 2.7-4   Top view of the Ergonomic Recommendations 2.7-32

Figure 2.7-5:   Side view of the Ergonomic Recommendations ... 2.7-33

Figure 2.8-1:   Examples of portable trestles 2.8-5

Figure 2.8-2:   View of driver removing a container from the
                 seventh tier 2.8-6

Figure 2.8-3:   View of driver placing container in delivery truck 2.8-6

Figure 2.8-4:   Two views of container product being transferred
                 to a residential freezer 2.8-11

Figure 2.8-5:   A side view of the load dock and the bumper of
                 the delivery vehicle 2.8-20

Figure 2.8-6:   Standard 10 inch channel lock pliers originally
                 used by employees to remove product from ovens 2.8-26

Figure 2.8-7:   New ergonomically correct hand tool introduced
                 to the pizza restaurant chain 2.8-26

Figure 2.8-8:   The handrails on the ladder pictured will have to
                 be installed after purchasing. The handrails
                 should be of light aluminum tubular
                 construction. Ladder weight without the
                 handrails is 23 pounds 2.8-31

Figure 2.8-9:   This figure denotes location of hydraulic unit on
                 the van. When unit is actuated, the arm will
                 lower to the side of the van at mid-chest height .. 2.8-32

Figure 2.8-10:   The cur rent ladder crane used should be
                 redesigned to incorporate a double-handle and
                 building stand to alleviate the amount of force
                 required to crank objects to the top of the
                 building 2.8-32

CHAPTER 1

INTRODUCTION

James Kohn and Celeste Winterberger

WHAT IS ERGONOMICS?

Most individuals know that ergonomics has something to do with people. They may not be aware of the fact that ergonomics examines human behavioral, psychological, and physiological capabilities and limitations. By understanding these capabilities and limitations, professionals in the field of ergonomics can then design new work environments, or modify established work environments, to maximize productivity, worker comfort and overall efficiency.

The primary objective of ergonomics is to improve human health, safety and performance through the application of sound people and workplace principles. The exciting feature of this approach is that it is a "win-win" situation where everyone benefits from the successful implementation of the ergonomic process. You may be asking yourself, "why is this objective so exciting?" If you were a health and safety professional practicing during the early 1970's, you could understand the advantages of this approach.

During the early years of the occupational health and safety movement in the United States, practitioners were often viewed as police officers. Their role was to enforce rules and regulations that often were viewed as interfering with the primary organizational objective of production. Recent history has revealed, however, that many organizations that have implemented ergonomic programs have observed a legal, moral, and financial advantage. Testimony by ergonomists indicate that they are actually being sought out by production supervisors because of the accomplishments and benefits that they have achieved in other departments. These benefits include decreases in injuries, absenteeism, complaints, and grievances experienced by the workers. These benefits also include increased productivity, decreased down time, improved materials handling and product flow.

For many organizations ergonomics can work hand-in-hand with a quality control program. By improving worker performance and production processes, superior products can be manufactured with less waste and fewer defects. As you can see, effectively establishing an ergonomic process in an organization can reap many benefits.

Terminology

Ergonomics is the discipline that examines the capabilities and limitations of people. The term ergonomics is based upon two Greek words: ergos meaning "work," and nomos meaning "the study of" or "the principles of." In other words, ergonomics refers to "the laws of work." The goal of ergonomics is normal to design the workplace to conform with the physiological, psychological, and behavioral capabilities of workers.

Most ergonomics problems arise out of pre-existing operations. It is then important for specially trained professionals to anticipate, recognize and identify ergonomic hazards. Evaluation and controls measures would be some of the activities that would have to be performed and implemented to eliminate ergonomic hazards and minimize ergonomic risk factors.

According to the OSHA meatpacking guidelines, "**ergonomic hazards** refer to workplace conditions that pose a biomechanical stress to the worker. Such hazardous workplace conditions include, but are not limited to, faulty work station layout, improper tools, excessive tool vibration, and job design problems. They are also referred to as (ergonomic) **stressors.**" The meatpacking guidelines defines **ergonomic risk factors** as "conditions of job, process, or operation that contribute to the risk of developing CTDs. Examples include repetitiveness of activity, force required, and awkwardness of posture." In addition, an **ergonomist** or **ergonomics professional** is defined in that same publication as "a person who possesses a recognized degree or professional credentials in ergonomics or a closely allied field (such as human factors engineering), and has demonstrated, through knowledge and experience, the ability to identify and recommend effective means of correction for ergonomic hazards in the workplace."

WHY IS ERGONOMICS IMPORTANT?

According to the Bureau of Labor Statistic's (BLS) report titled *Annual Occupational Injury/Illness Survey: Workplace Injuries and Illnesses in 1994,* the work related injury/illness frequency rate has declined steadily during the period between 1992 and 1994. The Bureau of Labor reported that in 1992 the injury/illness rate was 8.9 cases per 100 full-time workers. In 1993 the rate had declined to 8.5 cases and again in 1994 the injury/illness rate further decreased to 8.4 cases per 100 full-time workers. While the overall occupational health and safety statistics appear to be decelerating, ergonomic related incidences were found to be actually accelerating. Ergonomic incidents in the form of repeated trauma increased by more than 15 percent between 1992 and 1994 (Bureau of Labor Statistics, 1995). Considering Bureau of Labor data, it is obvious why ergonomics has been called the occupational injury/illness epidemic of the 1990s.

This ergonomic epidemic phenomenon, however, is not limited to the United States. Reports from Australia, Canada, Germany, New Zealand, Sweden, and the United Kingdom indicate that this is a global problem. According to a recent German medical journal study, Carpal Tunnel Syndrome was reported to be the most common compression syndrome in Germany accounting for almost 20 percent of all nerve lesions (CTDNews, 1995). New Zealand, for example, reported that in the year ending March 31, 1989, over $16.5 million in compensation had been paid to 6,200 recipients filing ergonomic related claims. The Ontario Workers Compensation Board reports that soft-tissue musculoskeletal disorders represented one-third of the disabilities among Ontario construction workers and accounted for two-thirds of their compensation costs (MacKinnon, 1995).

Health and safety professionals are aware of the growing magnitude of the ergonomic problem. In response to the ergonomic epidemic, some state and federal agencies have proposed the enactment of ergonomic legislation. For example, California has proposed legislation to address repetitive motion injuries in the occupational environment. California Title 8, General Industry Safety Orders, Article 106, Section 5110, was proposed as a result of the California Division of Labor Statistics and Research study of repetitive motion injuries. This study indicated that over 28,000 employers reported disorders associated with repetitive trauma. The federal government has not ignored the ergonomic problem. The proposed ergonomic standard has been in draft form for several years, but the moratorium on all new federal safety and health legislation has detained its passage. This standard was intended to address the repetitive motion injury problem as well as lifting and vibration issues.

Regulators are aware of the ergonomics problem. Health and safety professionals are also aware of the ergonomic epidemic. Even the public has been made aware of these issues through newspaper and television articles on the topic. Professionals and laypersons all agree that ergonomics is an issue that must be addressed.

Ergonomics has become an occupational problem of major proportions. National and international data point to ergonomic injuries and illnesses as the primary health issue of the 1990s. The medical and legal costs associated with ergonomically related dysfunctions are spiraling with no apparent end in sight. Ergonomic costs to employers are rising and are impacting their ability to compete in the global marketplace. An examination of ergonomic incidents and related costs seems appropriate.

HOW BIG A PROBLEM IS ERGONOMICS?

The Bureau of Labor Statistics (BLS) published 1994 workplace injury and illness statistics on December 15, 1995. This annual survey provided estimates of the frequency and associated incident rates of workplace injuries and illnesses based on OSHA 200 logs maintained by employers and submitted to BLS. The BLS survey indicated that 65 percent of all illness in 1994 were disorders associated with repeated trauma. Approximately 332,100 new repetitive motion cases were reported that year resulting in an incidence rate of 0.41 cases per 100 fulltime workers. BLS indicated that this was an increase of almost 10 percent over 1993 statistics and more than a 15 percent increase over 1992 survey results. According to this survey, "the private industries with the highest incidence rates of disorders associated with repeated trauma in 1994 are meat packing plants (12.6 cases per 100 fulltime workers), knit underwear mills (10.53 cases per 100 fulltime workers), motor vehicles and car body plants (9.6 cases per 100 fulltime workers), and poultry slaughtering and processing plants (8.3 cases per 100 fulltime workers)." Refer to Table 1-1 for the top 15 private industries with the highest incidence rates of disorders associated with repeated trauma in 1994.

Between 1990 and 1993, repetitive motion illnesses have increased by 63.1 percent in the United States. During that same period compensation cases have increased by 27.6 percent in Texas, 11.6 percent in Washington State, 16.2 percent in British Columbia, Canada, and 25.4 percent in Ontario, Canada (CTDNews, 1995). Official Swedish statistics (National Board of Occupational Safety and Health, 1993) revealed that the hands are the body part most affected by occupational injury (30 percent of all cases reported). Almost 71 percent of the occupational diseases reported in Sweden were musculoskeletal in nature. Occupational diseases to the hand-and-wrist resulted in an average of approximately 60 workdays lost, except for older workers who reported more workdays lost on average.

Industry	SIC Code	Incident Rate (per 100 fulltime workers)
Meat Packing Plants	2011	12.6
Knit Underwear Mills	2254	10.6
Motor Vehicles and Car Bodies	3711	9.6
Meat Products	201	8.8
Poultry Slaughtering and Processing	2015	8.3
House Slippers	3142	7.3
Motor Vehicles and Equipment	371	5.6
Motorcycles, bicycles, and parts	375	5.3
Men's and Boy's Underwear and Nightwear	2322	5.0
Engine Electrical Equipment	3894	4.8
Potato Chips and Similar Snacks	2096	4.6
Automotive Stampings	3465	3.8
Household Refrigerators and Freezers	3632	3.8
Men's and Boy's Work Clothing	2326	3.6
Vehicular Lighting Equipment	3647	3.6

Table 1-1: The 15 private industries with the highest repeated trauma incident rates with SIC Codes for 1994 (Bureau of Labor Statistics, 1995).

Specific industries are especially affected by ergonomic problems. For example, sprains and strains caused 37.6 percent of all lost-workday injuries in construction. One-quarter of those injuries affected the back (Dessoff, 1996). In general, 31.8 percent of all worker compensation costs in the United States are attributed to back injuries (National Council on Compensation Insurers, 1992).

Medical costs

In a recent publication, it was reported that more than one-third of all worker compensation costs, over $10 billion annually, goes to Cumulative Trauma Disorder cases (Banham, 1994). It is estimated that between 20 and 50 billion dollars are spent annually for back injuries. The American Academy of Orthopedic Surgeons has also estimated that cumulative trauma injuries have totaled $27 billion yearly in medical bills and lost workdays (Lin and Ciccone, 1994). Dr. Steven J. Barrer estimated that the average company with high repetitive motion risks would spend approximately $25,000 per carpal tunnel syndrome case (Barrer, 1991). NIOSH estimates that the average carpal tunnel case costs about $3,000 in benefits and

$40,000 in medical costs (NIOSH, 1989). The typical cost of carpal tunnel surgery is approximately $18,000 (Ramsey, 1995). Back injuries have been reported to cost an average of $9,000 in workers compensation and medical expenses (Dessoff, 1996). Medical expenses and worker compensation are expected to rise during the next decade. Ergonomic injuries and illnesses are expected to result in a greater proportion of a company's health expenses.

Legal costs

Companies may incur ergonomically related legal costs in a variety of ways. If an organization experiences a high frequency of repetitive motion or lifting related injuries, it could be cited by OSHA under the general duty clause. Citations could then result in substantial fines depending upon a variety of factors including the size of the company, the number of violations cited, an employer's willful disregard of employee safety or the good faith effort an employer demonstrates in attempting to abate ergonomic hazards. General Motors, for example, recently agreed to pay $420,000 in penalties to resolve ergonomic related citations associated with repeated motion and overexertion problems (Sand, 1995). Lehman Brothers, Inc., a much smaller company, was fined $4,650 for alleged word processing related wrist, back, and neck injuries in its New York office (Thornburg, 1994). Other companies cited for alleged ergonomic related violations include IBP Inc. ($3.1 million), Pepperidge Farm Inc. ($1 million) and a United Parcel Post Facility ($140,000).

The more common legal costs associated with ergonomic injuries is associated with employee lawsuits. CTDNews reported that a Long Island Railroad ticket clerk was awarded $55,000 for an alleged repetitive motion injury (1995). The injury was claimed to have developed while using a manual ticket vending machine. (Ryan v. Long Island et al., No. 92 Civ. 3029) (EDNY). The jury initially awarded the plaintiff $100,000, but reduced the award claiming that she was partially responsible for her injury. In another ergonomically related case, a former medical director presented arguments to the U.S. Court of Appeals for the Seventh Circuit (Chicago), claiming his former employer failed to accommodate his disability, Carpal Tunnel Syndrome. Filing under the Americans with Disabilities Act, the plaintiff claimed that the company fired him because he could not perform the essential functions of his job. No decision has been made to date.

Costs to organizations

In addition to the various ergonomic related costs already mentioned, organizations may experience losses associated with poorer productivity and indirect costs associated with all injuries and illnesses such as staff investigation time and increased absenteeism. Ergonomic programs have repeatedly been pointed out to make good economic sense (Ramsey, 1995; Rowan and Wright, 1994; Carson, 1994). Advantages that justify the implementation of an ergonomic program include: a reduction in the number of errors caused by poor working conditions, lower absenteeism and employee turnover, and improved overall productivity (Ramsey, 1995).

An example of lost productivity associated with ergonomic problems is the loss of a computer programmer as a result of repetitive motion trauma. The programmer could be on disability for six months or more resulting in project delays. In addition, the company may have to hire less competent temporary workers to replace the company trained programmer resulting in decreased productivity (Fine, 1995).

Another example of stealth costs to companies is manual materials handling. It is estimated that more than 60 percent of an employee's time is spent in material handling (Petersen, 1993). If product flow and handling could be improved, more time could be spent in production related activities. An example of improved productivity associated with reduced material handling is the assembly of medical components. Through automation of the assembly and conveyor systems one company realized a savings of approximately $100,000 and a 2.3 month payback period for the initial financial investment (Davies, 1995). Automation had eliminated tasks that had resulted in over $95,000 in worker's compensation medical expenses and increased productivity at the same time.

Benefits

According to a recent survey, companies are observing several benefits when implementing an ergonomics program (Kohn, 1996). Benefits most frequently cited included improved worker morale, improved productivity, better ergonomic hazard control, fewer complaints, fewer injuries reported, and improved health awareness. Brown has reported an 85 percent increase in productivity with a cost-benefit ratio of 1 to 10 with the implementation of ergonomic programs (Brown, 1991). Brown pointed out that the combination of increased production along with the reduction of worker compensation expenses reaped significant rewards when compared to the minimal costs incurred during the redesign of the workplace equipment and facility. Webb has also reported substantial gains in productivity as a result of modest investments in ergonomic programs and equipment (Webb, 1989). Numerous benefits have been demonstrated with the implementation of ergonomic programs (Refer to Table 1-2 for a list of these benefits of ergonomic programs). These benefits can positively impact both employees and employers.

Benefits of Ergonomic Programs
• Decreased errors and product defects
• Decreased time required to perform tasks (as a result of increased visibility and easier handling of stock, finished product, and tools)
• Reduced training and associated costs (as a result of a more stable workforce)
• Reduced hidden costs (such as disability salaries and insurance premiums)
• Improved morale
• Reduced worker discomfort
• Reduced fatigue related costs
• Improved hazard identification and control
• Improved quality
• Improved organizational performance
• Decreased loss of customers
• Improved company efficiency resulting from smaller workforce (resulting from reduced absenteeism and lost time injuries/ illnesses)
• Reduced management and supervision costs (fewer incident investigations and time spent solving related issues)
• Increased labor pool (older, less fit, and disabled workers can be employed in ergonomically designed workplaces)
• Decreased ergonomically related litigation
• Reduced disruption of work teams
• Reduced productivity fluctuations resulting from late-shift or late-week operations
• Improved general health awareness
• Increased occupational health and safety awareness
• Financial savings

Table 1-2: The benefits of implementing ergonomic programs.

Ergonomic areas of importance

When examining the human element associated with ergonomics, there are four broad areas of concern: physiological factors, psychological factors, behavioral factors and psychosocial factors. Physiological factors include anthropometric and biomechanical variables that influence an individual's ability to perform work related tasks. Anthropometrics is the study of human physical dimensions such as height, forward arm reach, or eye height in the sitting position. Biomechanics be defined as the study of the mechanical operation of the human body (Kohn, et al., 1996). It is the science of motion and force in living organisms.

Through the application of anthropometric and biomechanical principles, it is possible to reduce the physiological stressors placed on the worker's body via the redesign of tools, equipment, and facilities. This, in turn, reduces the likelihood of strain and sprain injuries prevalent in poorly designed occupational environments. The physiological factors of the ergonomic model will be discussed in greater detail later in this chapter. Psychological, behavioral, and psychosocial factors are the worker variables that contribute to various nonphysiological reactions that adversely affect worker performance. These will be discussed at this time.

Psychological

Psychology is the science that studies human behavior. Some of the psychological factors that have been found to contribute to ergonomic hazards in the occupational environment include: attention, memory, fear, boredom, fatigue, job satisfaction and occupational stress (Kohn, et al., 1996). An example of a psychological factor that can adversely influence worker performance and health would be fear. Individuals have been involved in traumatic accidents that have resulted in debilitating reactions not from the initial injuries sustained, but from the fear generated when those individuals returned to the accident scene. A master electrician with over 15 years of experience received second and third degree burns across his hands, arms and feet when he made contact with an energized pad mount transformer. He returned to work approximately six months later only to find that every time he entered the substation where the accident took place, he became very nervous. His hands would shake and he could not perform the work assignment. This example of a psychological ergonomic reaction points to health related dysfunctions that go beyond the frequently cited sprains and strains.

Behavioral

Behavioral factor refers to changes in worker activity that are observable and measurable. Reaction time, response accuracy and appropriateness, adaptation, and endurance are just a few examples represented by this category. An example of a behavioral factor might be response accuracy, reaction time and endurance of a hazard chemical spill clean-up worker during the hot temperatures of summer. While the psychological factor of fatigue and the physiological factor of exhaustion are readily apparent, behavioral factors under these conditions must also be considered. How quickly will an individual react to an unplanned change in the environment while using Level A (whole body, eye, and respiratory) personal protective equipment in temperatures approaching 100 degrees Fahrenheit? The environment and equipment used can adversely affect the worker's ability to react to an emergency or perform a complex sequence of tasks.

Psychosocial

Psychosocial factors are worker behaviors that are influenced by co-workers, supervisors, or the organization. It refers to worker behavior in a group environment (Kohn, et al., 1996). Concepts such as leadership style, employee motivation, organization reward systems, and attitude formation and change are just a few of the elements studied under this worker ergonomic category.

An example of a psychosocial factor could be the influence of co-workers versus management upon the use of personal protective equipment. A new employee wearing hearing protection could be in conflict with co-workers if the majority of the employees do not perceive a need for the protection. This could result in isolation and ridicule. There is less likelihood of the new employee following organizational guidelines for the use of personal protective equipment (PPE) under these conditions. If, on the other hand, new employees were trained in the hazards associated with their job, the company communicated its PPE policy, and co-workers urged compliance, there would be an increased use (behavioral factor of appropriate performance) of the required equipment.

WHAT ARE SOME STRATEGIES FOR IDENTIFYING
ERGONOMIC PROBLEMS?

There are a variety of tools and techniques used in the analysis of ergonomic hazards in the occupational environment. Anthropometric and biomechanical measurements are common starting points for the ergonomist (Carson, 1994). The health and safety professional would be concerned with the dimensions of workers as well as the dimensions of the workstations where tasks are performed. Observation of biomechanical body motions would then be conducted to determine if unnatural motions are required to perform required work activities.

Time and motion studies as well as behavioral sampling strategies are also of great importance in determining frequency, duration, force, and pace of the various motions. Quite often videotaping is required to closely study these variables. Job activity analysis, work task analysis, and work cycle analysis are just a few of the numerous methods that may be conducted in a thorough ergonomic risk assessment (Rogers, 1986).

Besides the methods listed above, specific activities should be analyzed. One example of a specific activity is lifting. Lifting could be analyzed using the NIOSH Lifting Guidelines. Surveys of employee opinions, health, and organizational activities are often a part of the analysis of ergonomic risk factors. The design of facilities and tools would also be analyzed to determine if they contribute to the ergonomic hazards identified in the workplace. Besides the processes, tools, equipment, facility, organizational variables and other possible factors, environmental stressors must also be studied.

Environmental monitoring of stressors such as temperature, lighting, noise, humidity, and air contaminants is another area of measurement required to obtain the ergonomic "big picture."

This lengthy list of factors for assessment of ergonomic problems reveals how complex the task is when starting to implement the ergonomic process in an organization. To increase the likelihood of success, organizations may wish to start ergonomic committees. While it is important to keep one person in charge of the process, an ergonomic committee can help spread the work load. Establishing ergonomic committees is not a simple task in itself. Organizations interested in establishing these committees should research the topic before getting started.

An in-depth review of all ergonomic analytical methods is not possible in this chapter. Readers are urged to review the references cited for more details concerning ergonomic methods of analysis.

HOW DOES HUMAN PHYSIOLOGY AFFECT ERGONOMICS?

It is very important for health and safety professionals to understand the physiological responses of the human body stressors in order to more effectively implement ergonomic solutions. These responses can be visible; however, most are invisible. Some of the more visible responses include sweating and/or bulging muscles. These outward manifestations, while important, do not represent an accurate picture of the stress placed on the human body. It is the invisible responses which give a much better picture of the overall functioning of the human body. This section will give the practitioner a basic overview of those physiological systems typically affected by ergonomic stressors.

the cell
the cardiovascular system
the nervous system
the musculoskeletal system

Table 1-3: The basic physiological systems typically affected by ergonomic stressors.

Following this discussion of basic physiology, musculoskeletal disorders and their effects on the physiological systems of the human body, as well as some common physiological disorders will be discussed.

Micro physiological response

Although the **cell** represents one of the smallest units of the body, it is at the root of all physiological responses. Each human being starts with one cell which then divides countless numbers of times to produce a human being. All cells are composed of essentially the same parts. However, it is important to note that most cells perform highly specialized functions within our separate organ systems.

The cell itself is composed of a cell wall. This cell wall performs two functions: it forms a protective layer between other cells and it allows for the transport of materials in and out of the cell. Proteins and protolipids are the most important elements found in the cell wall (Kutchai, 1988). If the cell wall is exposed to certain chemicals, it may break down and allow further transport of the hazardous material. Inside the cell are a number of structures (Freeman, 1982). Cytoplasm is the gel-like portion of the cell where the different structures exist. The nucleus contains the genetic material (called DNA) which is reproduced during the process of mitosis. Chemicals classified as mutagens or teratogens can alter the DNA of either the parent or the fetus.

endoplastic reticulum	contains RNA and ribosomes and plays an important part in cellular contractile systems including muscle contraction (Sherwood, 1993)
golgi apparatus	helps the cell in secretion and retention of protein materials (Freeman, 1982)
lysosomes	makes enzymes which break down excess material in the cytoplasm
mitochondria	produces the energy needed by the cell

Table 1-4: Several structures found in the cell.

The **cardiovascular system** is used to deliver blood, which contains nutrients and oxygen, to the cells. It also removes carbon dioxide and other waste materials from the cells. Carbon dioxide is expelled out of the body by the lungs while other waste materials are taken to the kidney for excretion from the body. This system consists primarily of the heart, lungs, veins, and arteries.

The heart is the engine of the body pumping blood through the arteries. First, the blood is taken to the lungs via the pulmonary artery. Oxygen is transferred into the red blood cells at the alveoli while carbon dioxide is passed off into the lungs for exhalation. Blood at this stage is bright red in color. Once the blood is taken to the cellular level, it transfers the oxygen and other nutrients to the cells and removes carbon dioxide and other waste products (at this point the blood is dark red in color). This blood is then returned to the heart and the process begins again. The most important thing for professionals to remember about the cardiovascular system is that it reaches every cell of the body allowing toxic materials full access to all of the body's organ systems (Berne and Levy, 1990).

There are two basic components of the **nervous system:** the central nervous system (CNS) and the peripheral nervous system (Willis, 1990). The CNS is composed of the brain and the spinal cord. In the brain, electrical impulses are sent to the various lobes.

cerebral cortex	where voluntary nervous impulses, sensory perception, and sophisticated mental events occur
thalamus	which is a relay station for synaptic input and has some role in motor control
hypothalamus	which controls the basic functions of the body
cerebrum	which controls muscle tone and muscle coordination for the more skilled activities
brain stem	which controls most of the autonomic nervous system (Sherwood, 1993)

Table 1-5: The lobes of the brain where electrical impulses are sent.

From the brain, nervous impulses are carried through the spinal cord for distribution to the rest of the body. The spinal cord is approximately 45 cm (18 inches) long and 2 cm (0.79 inches) in diameter. There are five different sections of spinal nerves. Eight cervical nerves can be found in the neck region. Twelve thoracic nerves can be found in the trunk area. Five lumbar nerves compose the lower portion of the spine. Finally, five sacral nerves make up the lowest portion of the spine. A coccygeal nerve can be found at the end of the spine which is often called the "tail bone."

To protect the highly fragile spinal cord, there is a bony structure composed of vertebrae. Each vertebra is separated from each other by cartilaginous structures called disks. The peripheral nervous system takes nervous impulses from the CNS and distributes them throughout the body. From there, the peripheral nervous system can be further divided into the afferent and efferent divisions. The afferent division carries information to the CNS on the status of body systems. Efferent division nerves carry nervous system commands to the different body systems.

There are two types of nervous system response. The first is a voluntary response. A voluntary response is caused when an afferent nerve fires in response to some external stimuli. For example, look at a simple wave from a friend. The eye will detect that someone is moving away which sends an impulse to the brain which will initiate a response by efferent nerves to the muscles of the arm to begin the wave function. In voluntary responses, the individual has control of the action.

There is a second type of nervous response which is controlled by what is known as the autonomic nervous system (ANS). For example, when an eye blinks, the individual does not have to think of moving the muscles which control that function. It just happens. The heart beats with a rhythm which is controlled by the nerves located in that organ.   When people exert themselves, the heart begins to beat faster but not because of a conscious effort made by that person.

Then, there are some responses known as reflex actions. Reflex actions are a learned response to a certain type of stimuli. can The jerking back of the finger when one touches a very hot object is one example of a simple reflex. When humans are born, these connections are not immediately made, but just let a child touch a hot object a few times and he will learn to move reflexively when they touch a hot object.

The **musculoskeletal system** consists of bones and muscles. Bones form the skeleton which protects organs and organ systems within the human body. The vertebra of the spine protects the spinal column, the ribs protect the heart and lungs, and the pelvis protects the reproductive and abdominal organs.

Bones also are used to support the weight of the human body. The bones of the legs and feet must be able to withstand the force of gravity and the constant pounding of walking, running and lifting. In order for humans to achieve movement, these bones must be connected by contractile tissue. This tissue is known as muscle.

Muscles consist of a number of fibers bundled together to form one mass. There are three types of muscles: voluntary, involuntary, and heart (cardiac muscle). Voluntary muscles are called skeletal or striated muscle and connect the bones of the body. Involuntary muscles are found in the internal organs and are known as smooth muscle.

It is important to note (Meiss, 1982) that muscles can only pull. Therefore, any muscle must have a restretch mechanism. In cardiac muscle, the restretch mechanism is accomplished by new blood entering the chambers, smooth muscles are restretched when material enters the organ and skeletal muscles work in pairs so as one muscle contracts the other is stretched. This arrangement of skeletal muscle pairs is known as antagonistic /protagonistic (for example biceps and triceps).

Skeletal muscle bundles are connected to the bone by tendons. The end of the muscle attached to the more stationary joint is called its origin while the attachment at the moving part is called its insertion. However, since humans are flexible there must be one additional mechanism available which allows for movement. These articulations are called joints. The bone and joint form a lever system with the bone being the lever and the joint being the fulcrum (Sherwood, 1993). There are six types of body movements which can occur around joints (refer to Table 1-6).

flexion	decreasing the angle of a joint such as bending the elbow
extension	increasing the angle of the joint such as straightening the leg at the knee
abduction	the moving of a body segment away from the midplane of the body such as raising the arm up and away from the side of the body
adduction	moving a body segment toward the midplane of the body such as lowering the arm from a horizontal position to a vertical position next to the body
rotation	moving a body segment in a circular motion around a joint such as "windmilling" the arm around the shoulder joint

Table 1-6: The six types of body movements which can occur around joints (Sanders and McCormick, 1993).

Fluid filled sacs called bursa (disks in the back) cushion the bones between the joints so the bones do not rub together causing pain. Ligaments stabilize the bones around a joint (Kroemer, Kroemer, and Kroemer-Elbert, 1990).

Macro physiological concerns

The effects of poor ergonomic design is typically not limited to one organ or organ system. Many times a number of systems will be affected, with the failure of one system frequently initiating the failure of another. Macro physiological disorders are a result of one or more physiological systems being impacted by a stressor(s). These will be covered by two broad categories: musculoskeletal disorders and physiological stress disorders.

WHAT ARE SOME ERGONOMICALLY RELATED INJURIES?

Most individuals recognize carpal tunnel syndrome and back injuries as the typical ergonomic injuries that are reported in the workplace. It is important to point out, however, that ergonomically related injuries include a wide variety of problems. These problems are often classified as cumulative trauma disorders or CTDs and tend to occur as a result of what has been termed "microtrauma" (Putz-Anderson, 1988). Microtraumas are the result of frequent repetitions of motions and forces associated with workplace tasks.

While CTD is one term used to describe this problem, there are many other terms that have been used to label the syndromes that result from microtraumas. Terms such as repetitive motion injuries, repetitive strain injuries, cumulative trauma disorders, and even occupational overuse syndrome (the term currently used in New Zealand) attempt to emphasize the chronic nature of these "illnesses." It is the frequent involvement of motion and forces, over a period of time, that characterizes the nature of typical ergonomic "illnesses."

Cumulative trauma disorders

Cumulative Trauma includes, but is not limited to, the following:

Back Region
- Herniated Disks: The weakening of the fluid filled disc that separates and cushions the vertebrae of the spine. This can result in the loss of internal fluid through the outer ring causing bone to bone abrasion or pinching of the spinal nerve.
- Ligament Sprain: Inflammation and swelling of the ligaments resulting from wrenching motions.
- Muscle Strain: Over use of the back muscles resulting in fatigue and pain. (Refer to Myalgia [Myofascial Syndrome] defined below).
- Mechanical Back Syndrome: General description of pain and discomfort in the back region.

Neck Region
- Tension Neck Syndrome: The inflammation of nerves and muscles in the neck region which results in pain and "stiffness."

Shoulder Region
- Thoracic Outlet Syndrome: The compression of nerve and blood vessels between the neck and the shoulder. It has symptoms

similar to that of carpal tunnel syndrome including numbness in the arm and fingers.

- Rotator Cuff Tendonitis: The inflammation of the tendons of the muscles attached to the rotator cuff area of the shoulder.

Arm, Wrist and Hand Region

- Epicondylitis: Also known as tennis elbow, this is an inflammation of the medial tendon of the elbow.
- Forearm Entrapment Syndrome (Pronator Teres Syndrome): Entrapment of the median nerve as it enters the forearm, often the result of inflammation of the muscles in this region.
- Myalgia (Myofascial Syndrome): "Pain in the muscle." This dysfunction is often the result of inadequate oxygen supply or physical damage.
- Radial Tunnel Syndrome: Pinching of the radial nerve by the muscles or bones in the epicondyle (elbow).
- Tendonitis: This occurs when the muscle and tendon is repeatedly tensed and the tendon becomes inflamed.
- Tenosynovitis: Caused by an excessive production of synovial fluid (a tendon lubricant); when the fluid accumulates, the synovial sheath will swell causing pain.
- Unsheathed Tendons: These are usually found in the elbow and shoulder when the joint is used for impact or throwing motions; the irritated tendons will then cause a shooting pain from the joint and down the arm.

- Carpal Tunnel Syndrome: The compression of the median nerve caused by the collapse of the tunnel containing the tendons, nerves, and blood supply to the hand.

- Cubital Tunnel Syndrome: Compression of the ulnar nerve in the epicondyle (elbow).
- Guyan Tunnel Syndrome: Ulnar nerve entrapment in the wrist.
- Digital Neuritis: Inflammation of the nerves in the hand.
- Trigger Finger: The tendon sheath surrounding the finger is so swollen that the tendon becomes locked causing jerking and snapping movements of the finger.
- DeQuervain's Disease: A form of stenosing tenosynovitis which affects the tendons on the side of the wrist and at the base of the thumb pulling the thumb back and away from the hand.
- Vibration Induced White Finger (Raynaud's Syndrome): This problem is caused by the use of vibrating hand tools in cold environments. In its early stages, vibration syndrome may cause numbness and tingling in the fingers but if it is allowed to

continue there will be a loss of all sensation and control in the hands and fingers.

- Ganglionic Cysts: Synovial fluid building up to form a bump under the skin.

Musculoskeletal disorders

Musculoskeletal disorders focus primarily on the skeletal muscles and their attachments to the bones. Since nerves play a major role in muscle contraction and feeling, they also have an important role in these types of disorders. There are three common forms of musculoskeletal disorders a health and safety professional should be concerned with: Cumulative Trauma Disorders of the extremities, back injuries, and segmental and whole body vibration injuries.

The most common of musculoskeletal disorders involve the back. Nearly 50 percent of all back injuries are caused by material handling (Sanders and McCormick, 1993). Of those material handling injuries, 50 percent are caused by lifting objects, 9 percent occur while pushing and pulling objects, and 6 percent occur while holding, wielding, throwing, or carrying objects. The following table shows data compiled on high risk industries for the most common incident types. Please note that injuries caused by overexertion (defined as incidents caused by maneuvering objects) have the highest rate of occurrence.

Event or Exposure	Industries	Rates
Overexertion	Nursing homes/Air transport	318/307
Struck by object	Logging/Wood containers	241/227
Fall on same level	Roofing/Water supply	121/118
Transportation incident	Taxicabs/School buses	114/102
Repetitive motion	Hats, Millinery/Men's suits	104/89
Assault by person	Residential care/Nursing homes	40/37

Table 1-7: Industries Having the Highest Incident Rates for Common Injuries per 10,000 Workers in 1994 from the Bureau of Labor Statistics (BLS), *Characteristics of Injuries and Illnesses Resulting in Absences From Work, 1994* (1996).

Often, during the lifting process, the back is used as a lever with the hips joints being the fulcrum. This is known as the stoop lifting method. Even using a proper lifting technique can create an enormous stress on the fifth lumbar (L5) and first sacral (S1) disk in the back. One factor known to increase the stress on the L5/S1 disk is the horizontal position of the load. It has been found that a simple inverse relationship exists between maximum weight lifted and the horizontal position of the load.

This can be calculated by using the following formula:

$$W = K (1/H)$$

where		
	$W =$	the maximum weight lifted
	$H =$	horizontal location of the load
	$K =$	a constant which depends on gender and vertical location of the load

According to Sanders and McCormick (1993) and NIOSH (1981), an object which normally weighs 44 pounds and is held eight inches from the body exerts a force of 400 pounds on the L5/S1 disc. Increase the distance to 30 inches and the force becomes 750. Muscles in the back and abdomen are also affected.

Some workers who have been performing lifting tasks for a number of years may even experience a herniated or "slipped" disk (Bridger, 1995). This injury will require surgical intervention resulting in an employee who is often partially disabled and under medical care for an extended period of time. Pre-employment screening will lower the initial risk to the employer. However, this does not absolve the employer from training employees in proper lifting techniques (straight back/bent knee lifts) and designing workstations that limit the risk of back injury.

There are a number of sources of anthropometric data for correct lifting heights. In addition, the revised NIOSH lifting equation (1991) can be used to determine the recommended weight limit (RWL) for manual material handling tasks. To calculate the RWL for a task, one would use the following equation:

$$RWL = (LC)(HM)(VM)(DM)(AM)(FM)(CM)$$

where   RWL = recommended weight limit

LC = load constant which is 51 lb

HM = horizontal multiplier calculated as 10/H where H is the horizontal of the hands from the midpoint between the ankles

VM = vertical multiplier calculated as [1-(0.0075|V-30|)] where V is the vertical location of the hands from the floor; measured at the origin and destination of the lift

DM = distance multiplier calculated as [0.82 + (1.8/D)] where D is the vertical travel distance between the origin and the destination of the lift

AM = asymmetric multiplier calculated as [1-(0.0032A)] where A is the angle of symmetry, the angular displacement of the load from the sagittal plane; measure at the origin and destination of lift

FM = Frequency multiplier as determined from NIOSH table where F is the average frequency rate of lifting measured in lifts/min; duration is defined to be: £ 1 hour; £ 2 hours; or £ 8 hours assuming appropriate recovery allowances

CM = Coupling multiplier as determined from NIOSH table

Using this formula and the following data, we can determine the RWL for a lifting task:

horizontal distance (H) = 6 inches
vertical distance (V) = 0 inches
distance (D) = 70 inches
angle (A) = 30°
frequency (F) = 0.5 lifts/min
coupling multiplier = 1 (assume good coupling)

LC = 51 lbs
HM = 10/H = 10/6 = **1.67**
VM = [1 - (0.0075|V-30|)] = [1 - (0.0075|0-30|)] = [1 - (0.0075(30)] =
[1 - 0.225] = **0.775**
DM = [0.82 + (1.8/D)] = [0.82 + (1.8/70)] = [0.82 + 0.0257)] = **0.846**

AM = [1 - (0.0032A)] = [1 - (0.0032(30))] = [1 - 0.096] = **0.904**
FM = 0.5 lifts per minute for 8 hours with a V of 70 from the table = **0.81**
CM = **1.0**

RWL = (LC)(HM)(VM)(DM)(AM)(FM)(CM)
RWL = (51)(1.67)(0.775)(0.846)(0.904)(0.81)(1)
RWL = 40.9 lbs

Finally, research on the use of back belts has not found any benefit to their use. In Autumn 1992, the National Institute for Occupational Safety (NIOSH) formed a working group to review the literature regarding back belts and their use. Its mission was to find any data which supported the premise that back belts reduce injuries. Because of the lack of scientific data regarding the efficacy of using back belts, the NIOSH working group concluded (Sweeney et al., 1994) that back belts do not prevent lifting injuries and do not reduce the hazards to employees caused by repeated lifting, pushing, pulling, twisting or bending.

Cumulative trauma disorders (CTDs) or **repetitive stress disorders** (RSDs) are a growing problem in the workplace (Refer to the list of CTDs found at the beginning of this section). According to Bureau of Labor Statistics, CTDs accounted for 6.8 percent of the nonfatal injuries and illnesses in U.S. businesses during 1994 (Bureau of Labor, 1995). This was up from 4.5 percent in 1993. Most of these injuries occur in the manufacturing sector. Cumulative trauma can be defined as injuries that are the result of repeated mechanical stresses (Putz-Anderson, 1992).

Pain
Restriction of joint movement
Soft tissue swelling
Loss of feeling
Reduction in manual dexterity

Table 1-8: Typical symptoms of cumulative trauma.

There are three categories of CTDs: tendon disorders, nerve disorders, and neurovascular disorders.

The most famous type of nerve disorder is carpal tunnel syndrome. This occurs when the tunnel containing the tendons, nerves, and blood supply to the hand is collapsed by repeated pressure to the underside of the wrist. After repeated compression of the carpal tunnel, the median nerve becomes compressed resulting in pain, numbness, and tingling in the hand. Eventually, if the individual continues to perform the same repetitive task, permanent loss of hand function may occur.

The most common neurovascular disorder is thoracic outlet syndrome. This is caused by the compression of nerve and blood vessels between the neck and the shoulder. It has symptoms similar to that of carpal tunnel including numbness in the arm and finger.

Vibration syndrome, Raynaud's syndrome or "white finger," is caused by the use of vibrating hand tools in cold environments. In its early stages, vibration syndrome may cause numbness and tingling in the fingers but if it is allowed to continue there will be a loss of all sensation and control in the hands and fingers.

About eight million individuals are subjected to some sort of vibration on a regular basis (Sanders, 1995). The whole body vibration frequencies of most importance to the industrial hygienist are from 0.1 Hz to 20Hz with accelerations of from 0.2 to 4 g (Eastman Kodak Company, 1993). These are typically caused by motors, compressor, and the impact of uneven elements of road/track (Tsimberov, 1994). Segmental vibration frequencies of importance range from 8 to 500 Hz with accelerations of 1.5 to 80 g (Eastman Kodak Company, 1993).

Typically, segmental vibration is caused by hand-held power instruments such as chain saws, jackhammers, drills, and torque guns (Tsimberov, 1994). Additionally, when safety professionals deal with vibration, they must understand the concepts of resonance and damping. Resonance occurs at a frequency where an object will vibrate at its maximum amplitude which is greater than the amplitude of the original vibration. In essence, the object will act as a magnifier at resonance frequencies.

Each substance has its own unique resonance frequency. "Since the body members and organs have different resonant frequencies, and since they are not attached rigidly to the body structure, they tend to vibrate at different frequencies" (Sanders and McCormick, 1993). Generally, the larger the mass of an object, the lower its resonance frequency. Damping is a natural or artificial method used to absorb vibration. A good example of a natural damping is when an individual uses the bending or straightening of his legs in response to movements within 1 Hz to 6 Hz range. One artificial method to use control vibration exposure in the workplace is to have the operator stand on a foam pad.

The Eastman Kodak Company (1993) states that there are three forms of vibration which the safety professional should be concerned with:

whole-body vibration	These are caused by transportation vehicles traveling over rough roads which create vibrations that are primarily in the vertical plane.
whole-body vibration	These are caused by production machinery. The vibration characteristics of equipment are based on the individual machine and should be measured by accelerometers.
segmental vibration	This is associated with the use of hand tools. It occurs primarily in the arms and hands of an individual and is based on the type of power tool being used.

Whole body vibration is typically concerned with frequencies below 100 Hz while segmental vibration occurs in the range of 40 Hz to 1000 Hz (Tsimberov, 1994). Although there is some overlap of frequencies between the two forms, the major difference between the two is that segmental vibration causes problems with specific body parts such as fingers, wrists, elbows, shoulders, or back, while whole body vibration deals with the effects on the entire human system. Over one million individuals are exposed to segmental vibration each year (Sanders, 1995). Some of the injuries which have been associated with segmental vibration include:

- small areas of decalcification seen in the X-rays of the small bones of the hand
- injuries to the soft tissues of the hand
- Osteoarthritis of the joints of the arms (Tsimberov, 1994)

The effects of whole body vibration vary depending on several parameters. Specific effects are discussed relative to a range of frequency, although there appear to be specific peak frequencies that can lead to resonance effects. The type of physiological effects of vibration on a given individual are primarily determined by the intensity, frequency and/or duration of exposure. In the 2–20 Hz range at 1 g acceleration, examples of the physiological effects of low frequency vibration include:

Loss of equilibrium
Chest pain

Abdominal pain
Shortness of breath
Nausea
Muscle contractions (Eastman Kodak, 1983)

Physiological stress disorders

While most of the research available seems to key on repetitive stress disorders such as carpal tunnel syndrome, there are other long term physiological disorders which are associated with on-going emotional stress. For example, unrealistic work schedules may cause stress for employees who are unable to keep up or employees could be stressed when they are asked to perform delicate assembly tasks in low light conditions.

It is well known that stress causes problems with the gastrointestinal system. Irritable bowel syndrome or chronic diarrhea can be caused by stress. If these symptoms continue, they could have serious consequences including the possibility of cancer. Another portion of the gastrointestinal system, the stomach, can also be affected by stress. When individuals are subjected to stress, the stomach produces excess stomach acid, which over time, causes a deterioration of the stomach lining. This often results in a sore called an ulcer. If the ulcer breaks completely through the stomach lining it is called a perforated ulcer which requires immediate surgical intervention.

The deleterious effects of stress can also be found in the cardiovascular system. Heart attacks are the major cause of death for many Americans. One of the contributing factors in heart attacks is stress. When an individual is subjected to stress, adrenaline, a powerful stimulant, is released by the adrenal glands. Adrenaline causes an increase in respiration and heart rate which can result in muscle tension. For short periods of time, this can cause an individual to be more productive. However, as the heart rate increases, so does the blood pressure. This is the pressure at which blood is pumped through the body. Chronic high blood pressure, if unchecked, can lead to heart attacks.

These are only a few of the occupational stress disorders which can affect people in the workplace. As medical research continues in this area, it will surely uncover other physiological disorders that are caused by on-the-job stress. Ergonomically designing workstations for the individual can reduce, if not eliminate, some of the causes of work related stress.

Causes of cumulative trauma disorders

Disorders such as carpal tunnel syndrome, tendinitis, tenosynovitis, DeQuervain's syndrome, and thoracic outlet syndrome are just a few of the many dysfunctions that rarely result from single incidents. These problems tend to occur as a result of what has been termed "microtrauma" (Putz-Anderson, 1988). Microtraumas are the result of frequent repetitions of motions and forces associated with workplace manipulative tasks. Many terms have been used to describe the syndromes that result from microtraumas. Terms such as repetitive motion injuries, repetitive strain injuries, cumulative trauma disorders, and even occupational overuse syndrome (the term currently used in New Zealand) attempt to emphasize the chronic nature of these "illnesses."

Physicians admit that there are numerous questions associated with the development of cumulative trauma disorders (Figura, 1995). A factor that compounds the difficulty of determining a cause is that there are just as many non-occupational causes of repetitive motion problems as there are occupational related causes. Repetitive motion problems may be caused by non-occupational disease factors including those given by Owensby in 1993:

> **System diseases:** Rheumatoid arthritis, acromegaly, gout, diabetes, myxoedema, ganglion formation and certain forms of cancer.
> **Congenital defects:** Bony protrusions into the carpal tunnel, anomalous muscles extending into or originating in the carpal tunnel and shape of the median nerve.
> **Tunnel size:** ... resulting from anthropometric ranges in wrist size.
> **Gynecological state:** Pregnancy, use of oral contraceptives, menopause and gynecological surgery.

Occupationally related causes result when workers are required to perform tasks that involve forceful and repetitive motions. Synergistic factors such as demanding awkward postures and unnatural positions of the upper extremities compound the risks. In addition, the lack of muscle strength recovery time may also contribute to the microtrauma. Workers involved in these types of tasks will often experience symptoms such as tightness, stiffness, or pain which can start in the fingers, hands, wrists, forearms, and elbows.

These symptoms can progress to sensations of tingling and numbness in the hands. The loss of strength and coordination along with radiating pain associated with inflammation of the median nerve can follow. These problems can be "productivity killers." 1993 BLS statistics reported that Carpal Tunnel Syndrome resulted in 30 median days away from work per case. Tendinitis, another repetitive motion syndrome, resulted in an average of 10 days away from work according to those same 1993 BLS statistics (Figura, 1995).

ARE SOME INDUSTRIES MORE LIKELY TO HAVE ERGONOMIC INJURIES?

Certain occupational environments have increased repetitive motion risks. Every workplace has unique tools or equipment that contributes to the ergonomic problem. A review of these conditions in some of these industries or locations follows.

Meatpacking

As cited earlier in the chapter, the BLS reported that in 1994 meat packing plants experienced 12.6 cases of carpal tunnel syndrome (CTS) per 100 fulltime workers. This significant incident rate has remained stable over the past several years and is responsible for the Occupational Safety and Health Administration publishing the document titled *Ergonomic Program Management Guidelines for Meatpacking Plants* (OSHA, 1991).

Lifting and twisting activities associated with material and product handling
Use of cutting tools such as knives
Power tools and their triggers
Vibration associated with power tools

Table 1-9: Ergonomic repetitive motion problems identified in the OSHA *Ergonomic Program Management Guidelines for Meatpacking Plants*.

The repetitive motion problem with the use of knives was associated with the handle design. Most meatpacking employees used a straight handled knife. As the carcasses went by on an overhead conveyor, employees made vertical slices on the carcass. The motion required by these slices caused the wrist to bend, increasing the likelihood of pinching the median nerve.

In addition, employees were required to grasp the knives tightly for the duration of the shift. To reduce the ergonomic risk found in this task, a pistol grip handled knife was introduced into the workplace. This allowed the employees to keep their wrists straight while performing the cutting task. In addition, a band was added to the handle, permitting the employee to hold onto the knife without having to grasp tightly. This band allowed the employees to release their grip and relax the muscle and tendons of the hand when not performing the cutting task.

Postures, positions and work methods were also analyzed. Ergonomic problems identified included static loading of the arm and shoulder muscles (contraction of muscles to support the position of body components). One recommendation suggested to eliminate this problem was automation. Articulated arms and counterbalances were also introduced so tools could be suspended by mechanical means. This modification meant that the employee used less muscular strength and force to hold the tools. Power tools were also introduced into the workplace in place of manually operated tools to reduce the repetitive motions required to perform many of the cutting tasks.

Office environment

New office equipment has contributed to the rising numbers of repetitive motion problems. Microcomputers are the most common contributor to the cause of carpal tunnel syndrome. Other equipment contributing to the office ergonomic epidemic are adding machines, cash registers, and tele-communication switchboards that require repetitive finger extension for data entry. These are motions that are similar to the keystrokes required when using a computer. The examination of office ergonomics will focus upon the computer.

With the increasing number of computers being introduced into the office environment, industrial hygienists have witnessed a steady increase in the prevalence of repetitive motion related cases. By replacing the typewriter with the computer or word processor, office workers have witnessed the elimination of several tasks that served as "breaks" for the keystroke motions required when typing (Barrer, 1991). Inserting and removing typing paper, using the carriage return at the end of each line, using corrective fluids, and pausing to look up the spelling of words were activities that interrupted the keystroke typing activity. Using computers in the modern office environment can require workers to perform over 23,000 keystrokes in a single work period without hand motion breaks (Barrer, 1991).

Does the frequency of keystroke motions increase the risk of repetitive motion syndromes? If the answer to this question is yes, then one would expect that occupations requiring high frequencies of keystroke motions would have higher incidences of repetitive motion problems. A 1990 study conducted in California and funded by NIOSH found that cases of Carpal Tunnel Syndrome were reported in approximately 50 different occupations. In this same study, 23 percent of the cases were reported in administrative support occupations and 13 percent were from cashiers (Owensby, 1993).

Examination of the interface between people and workstations can provide the clues necessary to identify potential ergonomic problems. Correct typing posture and position, as well as reduced frequency, pace, and force have been repeatedly reported to be the ergonomic solutions to computer elicited repetitive motion problems (Bisesi and Kohn, 1995; Parker and Imbus, 1992; Rogers, 1986). Job observations should be conducted to ensure that hips and thighs, as well as forearms are parallel to the floor.

Wrists should also be kept in a straight and level neutral position. While actually typing, the worker's wrists should be straight and not supported. Furniture such as desks, computer monitor stands, and chairs should be adjusted to fit the anthropometric measurements of the worker. Refer to checklists for the evaluation of workstations and worker posture to determine if deviations are contributing to the ergonomic problem (Bisesi and Kohn, 1995; Kroemer, 1983).

Construction

Unlike other industries that have experienced upper extremity repetitive motion problems, the majority of the ergonomic hazards in the construction industry are related to manual materials handling. The BLS reported that only 1.5 percent of all lost workday illnesses in the construction industry were due to repetitive motion (Dressoff, 1996). However, construction was second, only to transportation as the industry with the highest incidence rates for sprains and strains resulting in lost-workday injuries. Ergonomic hazards are prevalent in the construction environment.

Some of the ergonomic hazards that exist in the construction industry include: static positions, repetitive motions, material handling, awkward postures, overhead work and exposure to vibrating tools or equipment (MacKinnon, 1995). These are commonly associated with a wide variety of tasks from site excavation to internal structure finishing, such as painting, carpet laying, and trimming.

For example, during site excavation activities the use of heavy earthmoving equipment poses several ergonomic hazards. Whole-body vibration while operating earthmoving equipment, uncomfortable postures resulting from poorly designed seats, and repetitive motion associated with body twisting and turning while operating vehicles in reverse are just a few hazards common in this environment (Dressoff, 1996).

During bricklaying activities, material handling is the primary ergonomic hazard of concern. A worker may lift up to 1,000 bricks during a typical day. This translates to between 6,600 to 8,800 pounds (about 3,000 to 4000 kilograms) daily. In addition, the bricklayer may perform 1,000 trunk-twist flexions during the course of a work shift (MacKinnon, 1995, Dressoff, 1996).

These activities have a substantial impact upon the cardiovascular, nervous and musculoskeletal systems. The cardiovascular system is under stress during this activity because oxygen and nutrients are being distributed by the heart and circulatory system to the muscle groups performing the work (Grandjean, 1988). The cellular metabolic process in these muscles converts the oxygen and nutrients into mechanical energy used during brick handling. As muscles work harder, the cardiovascular system must work harder to provide the much needed energy. If the worker is in poor health, cardiovascular problems can occur.

The nervous and musculoskeletal systems are also under substantial stress during material handling activities. During bricklaying trunk-twist flexions may result in pinched nerves in the L5/S1 area of the spinal column. Overexertion muscle injuries such as strains of the arm (triceps and brachioradialis), shoulder (deltoids and trapezius), and back muscles (latissimus dorsi, infraspinatus, teres minor and major) may occur.

Other studies have found that concrete-reinforcement workers experience high frequencies of back pain. This ergonomic problem results from spending significant portions of a shift bent over to tie reinforcement rods, commonly referred to as rebar (Dressoff, 1996). Electricians were also found to spend a considerable amount of time in awkward postures and positions while installing cable. However, upper extremity hazards associated with the use of pliers and screwdrivers also posed potential repetitive motion threats for these workers.

Low cost solutions have been recommended for many of the problems identified in the construction industry. First, evaluation of tasks must be performed to reduce muscular effort, as well as the frequency of lifting and climbing. In addition, repetitive motions observed should be reduced by way of work site modifications and the use of alternative tools. For example, the use of height-adjustable work platforms was found to reduce the frequency of bending for bricklayers (MacKinnon, 1995). Concrete mixers required to open and pour bags of mortar into a mixer could use an inexpensive stand to raise the height of the bags and reduce the frequency of bending (Dressoff, 1996). Smaller bags of product that weigh less is another low cost solution to ergonomic material handling problems in the construction industry.

Manufacturing/assembly

On April 26, 1994 the BLS released a flyer titled *Work Injuries And Illnesses By Selected Characteristics, 1992*. In this flyer repetitive motion characteristics were profiled. The report indicated that women accounted for two-thirds of the nearly 90,000 repetitive motion injuries and illnesses cases. This report indicated that 18 percent of the repetitive motion cases reported resulted from repetitive use of tools. Thirty-one percent of the repetitive motion cases reported resulted from placing, grasping, or moving objects other than tools. Approximately 56 percent of these cases occurred in manufacturing. Occupations most frequently cited were operators, fabricators and laborers (51 percent), machine operators (24 percent), and assemblers (8 percent).

As revealed in the above mentioned statistics, manufacturing operations have experienced significant losses associated with ergonomic hazards in the workplace. Manual material handling and repetitive motion are the two leading ergonomic causes in this environment. The material handling statistics are just as staggering as the repetitive motion statistics. In 1993, approximately 28 percent of the total estimated days away from work in this country, or 21,000,000 days, resulted from lifting injuries (National Safety Council, 1994). When viewed together, ergonomic related losses from material handling and repetitive motion cost companies billions of dollars every year.

Most jobs in the manufacturing environment involve manual material handling. Workers handle raw materials and/or finished products. They may also be required to carry tools and/or containers. In addition, packing operations go on continuously at most facilities.

To avoid material handling injuries many companies have employed the use of scissor lifts, carts, dollies and powered hand trucks. Palletizing materials and using forklifts are additional solutions to material handling problems.

Repetitive motion problems pose a greater challenge for the health and safety professional. The National Safety Council reports that 2,925,000 days are lost each year as a result of repetitive motion injuries (1994). The frequency of repetitive motions that can be tolerated without experiencing injury varies by age, gender, health and a variety of other factors. No specific safe repetitive motion threshold limit has been established to date. However, the proposed OSHA Ergonomic Standard recommended modification of the work process if one of the following attributes was observed:

1. tasks requiring over 2,000 manipulations per hour (such as keystrokes during word processing),
2. manual task work cycles that are 30 seconds or less in duration (such as pinch grip repetitive motions during small component assembly),
3. repetitive tasks which exceed half of the worker's shift (a data entry clerk who performs calculation keystrokes during 90% of the shift or inspectors who manipulate parts during inspection 85% of their shift).

The use of hand and power tools contribute to repetitive motion problems. Using a power grip to hold a file or wrist deviations associated with soldering are just some of the ergonomic hazards that have caused repetitive motion injuries. Applying labels to equipment manually or packaging finished products are additional causes of ergonomic upper extremity hazards.

Repetitive motion problems are also associated with the loading and unloading of power presses and riveting machines. For example, employees are often required to use a pinch grip (grasping a component between the thumb and forefinger or thumb and remaining fingers) when removing small components from a storage bin and placing the part into the press die. In addition to the repetitive pinch grip, dorsiflexion and palmar flexion may result during the component handling as well as during power press manual stock feeding and removal activities.

Wherever possible, modify the task or the process to eliminate the frequent repetitions, stressful postures, or muscle exertion requirements. For example, one company required that the finished product be lifted over the side and placed into a cardboard box for sealing and labeling. By reducing the height of the work table and placing the box on its side, workers could roll the product into the box. This eliminated the lifting motion as well as the bending of the workers' wrists.

At another location, product was removed from a conveyor at the end of an assembly line and placed on pallets that were on the shop floor. The company introduced a scissor lift that supported the pallets. This eliminated the need to bend when loading the finished product onto the pallet. The scissor lift is a mechanism that raises and lowers as stock is loaded or unloaded. In this way, the pallet height is maintained near knuckle height for the standing employee reducing the need to bend or extend overhead during the palletizing process.

WHAT ARE SOME METHODS TO ELIMINATE OR CONTROL ERGONOMIC PROBLEMS?

There are a number of ways that the ergonomic hazards discussed above can be mitigated. This section will discuss four different methods for controlling the ergonomic environment of a workplace. In addition, different industrial scenarios for each method will be given.

Human control methods

The best human control method is proper training. For example, employees performing manual lifting tasks should be trained in proper lifting methods. Assuring that employees know how to lift properly can reduce the number of back injuries far better than using Personal Protective Equipment (PPE) such as back belts. Training on the proper use of special ergonomic equipment such as adjustable chairs can help ensure that employees actually use those features that make their job easier.

Another human control method is exercising prior to the beginning the work shift. Most individuals who workout know that stretching the muscles before strenuous exercise can lower the risk of injury. Many companies, including Wal Mart, have started using stretching exercises to prevent injuries to their employees during the work day. Some research has been done regarding the success of these exercise programs in reducing workplace injuries. In a review of eight research studies, McGorry and Courtney (1995) found that not enough data exists to support the contention that on-site exercise programs alone will reduce the number of CTDs in the workplace. They did conclude that studies of exercise programs coupled with engineering controls and/or administrative controls showed positive results. However, they suggest that more controlled studies be performed to determine whether exercise alone will be effective in reducing CTDs.

Wellness programs have also become very popular in the past few years. Many companies feel that promoting the physical fitness of their employees may reduce the number of injuries on the job. GlaxoWellcome, at their Zebulon, North Carolina manufacturing facility, has a fitness center on-site where employees can workout 24 hours a day. Health Maintenance Organizations (HMOs) such as Healthsource North Carolina, Inc., have offered special incentives to their subscribers who exercise on a regular basis. Some of these incentives include lower cost memberships at local health clubs, as well as rebates for the purchase of home exercise equipment.

Human environmental control methods

Whether it is temperature, noise, lighting and/or vibration, most of the environmental stressors at the workplace place significant demands on the physiological systems of the body. Methods for reducing the effects of each of these stressors will be discussed in this section.

Extremes in temperature cause problems with the regulation of the core temperature. Since the normal body core temperature is approximately 96.8 to 98.6 °F (Bridger, 1995), any temperature stressor that has the potential to raise or lower these values can be dangerous. One of the best ways to control the effects of temperature is to allow the employee's body to acclimate to the environment. For warm climates, allow the worker two weeks to adjust. This may be achieved by having the employee work in the warmer environment for increasing periods of time during the two week period.

There is doubt about humans acclimatizing to cold climates; however, the same practice of increasing work exposures over a defined period of time may allow workers to somewhat acclimate to the colder temperatures. Clothing can also be used to control temperature extremes. Insulated clothing, including gloves and hats, can be used in cold environments. An important concept to remember in cold environments is the use of natural fibers especially close to the skin. If the clothing near the body becomes saturated with sweat because the fibers do not allow the moisture to move away from the body, the individual may become cold even if wearing multiple layers of clothing.

For warm environments, loose, light colored clothing works best. However, some operations, such as arc welding, require the worker to wear special protective clothing. In these cases, cooling vests may be used to keep the body core temperature within normal limits.

Noise can be dangerous as well as annoying. High noise levels can cause permanent hearing damage. Noise may also be a barrier to effective communication. The best way to control environmental noise is to isolate the employee from the noise source. This may call for the use of sound attenuating barriers or PPE such as earplugs or muffs.

Inadequate or improper lighting can place an employee under stress. This includes either too little or too much light. Proper monitoring methods should be used to determine the level of illumination in a given area. If the illuminance is not at the correct level for the task, efforts should be made to adjust the illuminance levels. One way of increasing illumination is the use of task lighting. Other methods may involve the use of different light sources. This may include the use of halogen lights, sodium vapor lights, or fluorescent lights. Glare can also create stress for the employee. Efforts should be made to reduce the effects of glare especially on glass covered controls and computer workstations. Glare can be reduced by using diffused overhead lighting sources and task lighting.

Segmental vibration can be controlled by isolating the worker from the source. This can be done by using hand tools that are ergonomically designed or by providing the employee with gloves. Whole body vibration can be reduced by having the employee stand and/or sit on a cushioning material.

Equipment/facility control methods

Most ergonomics texts will indicate that engineering controls (equipment, workstation and facility design) are the best method of reducing or eliminating ergonomics disorders. There is an abundance of anthropometric data available on the design of equipment, workstations, and facilities. Van Cott and Kincade's text (1972) contains a wealth of anthropometric data regarding the design of equipment and controls. All military contractors are required to comply with the anthropometric data found in MIL-STD 1472D when they design equipment, workstations and facilities for use by U.S. military forces. NASA is also an excellent source for design data, most of which can be easily accessed using the World Wide Web. In addition, web sites, such as ErgoWeb (http:\\ergoweb.mech.utah.edu\) or CTDNews*Online* (http:\\CTDNews.com\) are other sources of information that can be used in design applications.

With all these sources of data and the continuing research on equipment, workstation, and facility design, it would seem that these control methods could be easily implemented. Unfortunately, this is not the case. With most industries trying to remain competitive by downsizing, economics begins to mean more than ergonomics.

Many tool, equipment, and furniture manufacturers now offer ergo-nomically designed products. One cannot open a safety or engineering related magazine or journal which doesn't have a least one advertisement for companies selling ergonomic products. Sears even carries some hand tools which have been designed for ergonomic use. Many web sites also exist for companies marketing ergonomic equipment. In addition, many consultants are offering their services to industry for the design and implementation of ergonomic solutions for equipment, workstations and facilities.

Organizational control methods

If design efforts fail to reduce ergonomic disorders, organizational control methods (sometimes referred to as administrative controls) are the second best control methods available. Most of the organization controls involve limiting the amount of time that a worker is exposed to environmental stressors which may include:

> providing rest breaks
> using job rotation
> reducing the work rate
> requiring the use of PPE

Organizations can also create a corporate environment where ergonomics is considered important to the overall success of the company. Workers can be encouraged to submit suggestions for solutions to ergonomic problems found either in their workstation or department. Successful implementation of these ideas would result in the employee sharing part of the first year's savings. The use of safety committees composed of employees from all levels is another way that organizations can encourage the use of ergonomics.

WHAT CAN BE LEARNED FROM THE EXPERIENCE OF OTHERS?

Many times professionals in an organization incorrectly see themselves as having to develop strategies and programs to eliminate problems in the workplace. They mistakenly believe that their problems are unique. The old saying, "Why reinvent the wheel?" comes to mind when situations like these arise. It is important to realize that if you have a particular ergonomic problem, in all likelihood, other organizations have had similar problems. That is part of the purpose for this book.

The intent of this book is to provide the reader with examples of ergonomic problems that health and safety professionals have faced in a wide variety of occupational environments. You may discover that an organization in an unrelated industry faced a similar problem to the ergonomic problem that you may be confronting. By examining the specific nature of the ergonomic problem and studying the strategies and recommendations for its modification, you may develop creative strategies for eliminating your ergonomic "challenge." The key to successful implementation of the ergonomic process is creativity.

Creative problem solving is the foundation for your organization's ergonomic system. What is creative problem solving? Creative problem-solving is concerned with applying information, tools, techniques, or methods in a new or different way. By looking at an ergonomic problem from a new or different perspective, creative solutions will evolve.

WHY CASE STUDIES?

Case studies can help you develop those creative solutions needed to solve ergonomic problems. The use of case studies for gaining insight and understanding of real world problems is as old as antiquity. The parables and other stories of the Bible could be considered as religious or philosophical case studies.

Case studies have been found to be an effective educational method that has been used by a variety of disciplines. Schools of business, law and medicine frequently rely on the use of case studies in their educational programs. The benefit of case studies is that they present situations that individuals have faced and show how those perplexing situations were resolved. This book can serve that same function no matter whether you are reading this publication for a class or as a resource to help you develop a broader and, perhaps, more creative solutions to your ergonomic problems.

HOW IS THE CASE STUDY BOOK ORGANIZED?

The case studies that appear in the following sections are organized by industry. The titles attempt to explain the specific ergonomic problem that was studied in that industry. These case studies may have different headings for the various sections that are presented. This is because this book is a compilation of numerous case studies from a wide variety of authors. All the case studies, however, have the same basic elements.

Each case study starts with a basic introduction to the industry or the particular problem of concern. This is followed by a brief statement of the specific ergonomic problem that will be examined. A short description of information related to facility characteristics, employees, job title information and other relevant data is then presented. The case study ends with a list of recommendations and/or solutions to eliminate or modify the ergonomic problem.

CONCLUSION

The following case studies are intended to serve as a model for your use. They can serve as examples that can be replicated to address your particular ergonomic problem. These case studies can also serve as a jumping board to help develop your own strategy for eliminating your specific ergonomic problem. They can also provide you with insights into unfamiliar occupational environments and the specific ergonomic problems that they have posed. No matter how you apply the information found in this resource, keep in mind that sharing experiences is the most effective way to learn about ergonomic problems and their solutions.

If, as a result of the case studies presented in this book, you develop successful approaches to solving an ergonomic problem, please consider sharing your knowledge and experiences. Future editions of this publication will include additional contributions. Please feel free to contact the publisher or author if you have a case study that would positively contribute to this work.

REFERENCES

Banham, R. (1994, May). The new risk in ergonomics solutions. *Risk Management*, 22-30.

Barrer, S. (1991, January). Gaining the upper hand on carpal tunnel syndrome. *Occupational Health & Safety*, 38-43.

Berne, R., & Levy, M. (1990). Cardiovascular system. In R. Berne & M. Levy (Eds.). *Principles of Physiology* (pp. 188-311). St. Louis, MO: Mosby.

Bisesi, M., & Kohn, J., (1995), *Industrial hygiene evaluation methods*. Boca Raton, FL: CRC/Lewis.

Bridger, R. (1995). *Introduction to ergonomics*. New York: McGraw-Hill.

Brown, R., Todd, G., & McMahan, P. (1991,). Ergonomics in the U.S. railroad industry. *Human Factors Bulletin,* 34,

Bureau of Labor Statistics. (1996, May 8). *Characteristics of injuries and illnesses resulting in absences from work, 1994* [On-line]. Available: ftp://stats.bls.gov/pub/news.release/osh2.txt.

Bureau of Labor Statistics. (1995, December 15*). Bureau of Labor Statistics annual occupational injury/illness survey: Workplace injuries and illnesses in 1994,* [On-line]. Available: http:// ergoweb.mech.utah. edu:80/Pub/info/bls.html.

Bureau of Labor Statistics. (1994, April 26). *Workplace injuries and illnesses by selected characteristics, 1992. Repetitive motion profile.* [On-linc]. Available: http://dragon.acadiau.ca/~rob/rsi/usflier.html.

Bureau of labor statistics releases 1994 repeated trauma numbers. (1995). *CTDNewsOnline*. [On-line]. Available:http:\\ctdnews.com\bls94.hmtl.

Carson, R. (1994, August). Key ergonomic tips for improving your work area design. *Occupational Hazards*, 43-46.

Davies, J. R. (1995, February). Automation and other strategies for compliance with OSHA ergonomics. *Industrial Engineering*, 48-51.

Dessoff, A. (1996, January). Seek simple solutions for ergonomics problems in construction. *Safety and Health*, 62-65.

Eastman Kodak Company. (1983). *Ergonomic design for people at work* (Vol.1). New York: Van Nostrand Reinhold.

Figura, S. (1995, November). Dissecting the CTS debate. *Occupational Hazards*, 28-32.

Fine, D. (1995, May). A break now saves money later. *Infoworld*, 54.

Freeman, A. (1982). Cellular function and fundamentals of physiology. In E. Selkurt (Ed.). *Basic physiology for the health sciences* (2nd Ed.) (pp. 3-29). Boston: Little, Brown, and Company.

Grandjean, E. (1988). *Fitting the task to the man*. Philadelphia: Taylor & Francis.

Kohn, J. (1996, April). Evaluating ergonomic progress. *Ergonomic News*.

Kohn, J., Friend, M., & Winterberger, C. (1996). *Fundamentals of Occupational Safety and Health.*, Rockville, MD: Government Institute Press.

Kroemer, K. (1983) *Ergonomics of VDT workplaces*. American Industrial Hygiene Association: Akron, OH.

Kroemer, K, Kroemer, H, & Kroemer-Elbert, K. (1990). *Engineering physiology: Bases of human factors/ergonomics* (2nd Ed.). New York: Van Nostrand Reinhold.

Kutchai, H. (1988). Cellular physiology. In R. Berne & M. Levy (Eds.). *Physiology* (2nd Ed.) (pp. 5-65). St. Louis, MO: Mosby.

MacKinnon, L. (1995, July/August). Construction: Building a safer industry. *OH&S Canada,*

McGorry, R. W. & Courtney, T. K. (1995). Exercise and cumulative trauma disorders: The jury is still out. *Professional Safety*, 40, 22-25.

Meiss, R. (1982). Muscle: Striated, smooth, and cardiac. In E. Selkurt (Ed.). *Basic physiology for the health sciences* (2nd Ed.) (pp. 3-29). Boston: Little, Brown, and Company.

National Council on Compensation Insurers. (1992*). Workers' compensation back claim study.* Boca Raton, FL: NCCI.

National Safety Council. (1994). *Accident facts* (1994 Ed.). Itasca, IL.

National Institute for Occupational Safety and Health. (1989, March). *Carpal tunnel syndrome: Selected references.* Washington: U.S. Department of Health and Human Services.

Occupational Safety and Health Administration. (1991). *Ergonomics program management guidelines for meatpacking plants.* Washington: U.S. Department of Labor.

Official Statistics of Sweden, National Board of Occupational Safety and Health. (1993). *Occupational diseases and occupational accidents* (1991 Ed.). Stockholm: National Board of Occupational Safety and Health.

Owensby, G. (1993, March/April). Carpal tunnel syndrome *Ohio Monitor*, 8-11.

Parker, K., and Imbus, H. (1992). *Cumulative trauma disorders: Current issues and ergonomic solutions-A systems approach.* Boca Raton, FL: Lewis Publishers.

Petersen, D. (1993, June). Streamline your workflow. *American Printer,* 18-21.

Putz-Anderson, V. (1988). *Cumulative trauma disorders: A manual for musculoskeletal diseases of the upper limbs.* Philadelphia: Taylor & Francis.

Ramsey, R. (1995, August). *What supervisors should know about ergonomics.* Supervision, 10-12.

Rowan, M. P., and Wright, P. C. (1994) Ergonomics is Good for Business. *Work Study*, Vol. 43, No. 8, 7-12.

Rogers, S. (1986). *Ergonomic design for people at work.* (Vol. 2). New York: Van Nostrand Reinhold.

Sand, R.H. (1995, Summer) "Firestone wins an Ergonomics Battle...," *Employee Relations Labor Journal.* Vol. 21, No. 1, 139-144.

Sanders, M. S., & McCormick, E. J. (1993). *Human factors in engineering and design* (7th ed.). New York: McGraw-Hill.

Sanders, R. E. (1995, May/June). Bad vibrations. *Workplace Ergonomics*, 1, 22-25, 28-29.

Sherwood, L. (1993). *Human physiology: From cells to systems* (2nd Ed.). Minneapolis/St. Paul, MN: West Publishing.

Sweeney, M., Gardener, L., Parker, J., Walters, T., Flesch, J., Huduck, S., & Smith, S. *Workplace use of back belts*. [On-line]. Available:http:\\ergoweb.mech.utah.edu\Pub\Info\Std\ backbelt.hmtl.

Thornburg, Linda (1994, October). Workplace ergonomics makes economic sense. *HR Magazine*, 58-59.

Tsimberov, D. (1994, October). Guidelines warn when the rattling can cause harm. *Workplace Ergonomics*, 1, 26-32.

Van Cott, H., & Kincade, R. (Eds.). (1972). *Human engineering guide to equipment design*. Washington: US Government Printing Office.

Webb, R. (1989). A feeding frenzy. *OH&S Canada*, 5, 91-92.

Willis, W, Jr. (1990). Nervous system. In R. Berne & M. Levy (Eds.). *Principles of Physiology* (pp. 56-151). St. Louis, MO:Mosby.

CHAPTER QUESTIONS

1. According to the Bureau of Labor Statistics 1994 survey, 65 percent of all illnesses were disorders associated with what occupational injury?

2. Identify the industry with the highest incidence rate of disorders associated with repeated trauma in 1994.

3. Provide examples of occupational legal costs associated with repetitive motion or lifting related injuries.

4. List what you consider to be the top five benefits associated with the implementation of ergonomic programs.

5. The study of human physical dimensions is a definition for which discipline?

6. Reaction time and endurance are examples of topics associated with what ergonomic domain?

7. Explain why time and motion studies and behavioral sampling are of great importance in ergonomic assessment?

8. Disk herniation associated with manual material handling and poor lifting methods occurs in what vertebrae area of the spinal column?

9. Explain how belt backs are viewed by ergonomists.

10. Carpal tunnel syndrome is a repeated trauma dysfunction associated with the compression of which nerve?

11. Numbness, blanching of fingers, loss of muscular strength and finger control resulting from vibrating equipment and aggravated by cold environments is a definition for which repeated trauma syndrome?

12. Identify the forms of vibration which are of ergonomic concern in the occupational environment and explain how they differ from one another.

13. List examples of non-occupational causes of repetitive motion problems.

14. List examples of controls that could be used to minimize exposure to ergonomic stressors.

15. List and describe administrative controls that can be used to address ergonomic problems.

CHAPTER 2.1

AGRICULTURE CASE STUDIES

When one thinks about ergonomics it is difficult not to think of the meatpacking industry. Meatpacking and rendering facilities were at the forefront of ergonomic hazard recognition as a result of the high incidence of repetitive motion problems. However, the rest of the agriculture industry is often neglected by many occupational health and safety professionals. Ergonomic problems are as prevalent on family farms, commercial farms and related agricultural industries such as agricultural co-operatives as in any manufacturing facility.

The following case studies examine equipment layout and design problems as well as the traditional overexertion injuries that can be experienced in any occupational environment. Are you employed in a nonagricultural industry? Compare the problems presented in the following case studies with those that you have observed. At the completion of this section you may wish to list the solutions and recommendations. Then consider how those solutions or related strategies might be applied to positively impact ergonomic hazards of concern to you.

ERGONOMIC JOB DESIGN IN THE HOG INDUSTRY

William C. Hilton

INTRODUCTION

Too often workplaces have been designed based on "efficient movement of product" or "best locations for machines." All of these workplace designs are developed with little thought given to how people fit in. Designers have expected people to adapt to whatever system has been devised. Unfortunately, the human body cannot adapt to everything. People are different and have limitations. People behave and react in certain ways that do not always fit into traditional management concepts of how work should be done. The idea of ergonomic job design is to understand the anatomy, physiology, and psychology of people and design for those factors. By using this approach, employee well-being and workplace efficiency can be obtained and/or optimized (Jacobs, Larson, and MacLeod, 1990).

There are three areas of importance in the ergonomic design of the workplace. They are:

- **Physical issues** — factors such as furniture, equipment, heights, reaches, mobility and good lighting.
- **Usability** — making the workplace user friendly or "adjustable."
- **Work organization** — issues such as responsibility, job-decision latitude, good work/rest schedules, and job rotations.

Use of these objectives, combined with a comprehensive ergonomic job analysis checklist, should minimize the ergonomic risk of work related injury (Davis, Grubbs, and Nelson, 1995).

CASE STUDY

On May 14, 1996, at 6:00 a.m., Kathy M., a farrowing assistant at a large hog farm, reported for work as usual. During the morning meeting, Kathy was informed that two other employees were going to be out sick for the day. This meant that she, along with the other employees, would have to perform the absent worker's normal daily duties in addition to their own duties. What this meant for Kathy was that she was going to have to pull the feed levers in the gestation houses at predetermined times during the day. This was a job she was unaccustomed to and a job that she should have never performed.

At approximately 6:45 a.m. Kathy approached feed line number four. She noticed that the feed levers at this station were different from the previous three she had already pulled. These feed levers were different because they had been installed improperly. (Note: refer to Figures 2.11-1 to 2.11-3 for layout of the feed levers.) Improvising, because all of the other employees were busy, she climbed approximately two feet above the cement floor in her size seven boots (she actually required a size five work boot). Now she had both feet on opposite gates; she was ready to pull a feed lever. This feed lever was different from all the rest. It required that Kathy push it, rather than pull it, to deliver the feed into the sows trough. As she stated in a letter that she later wrote, "It took all of my 5'0, 97 pound frame to push that first lever." While it was difficult to push the lever, she did complete the task.

She proceeded to push the second lever. However, this time she was not successful. As she got three-fourths through the turn she overextended and her hand slipped off the lever. This caused the lever to recoil and strike her under the chin. She immediately fell to the ground in pain and was rushed to the hospital emergency room. There she was referred to an oral surgeon, who treated her for a bilateral fractured mandible. In laymen terms she had her mouth wired shut for six weeks. Kathy did not return to work for over two months, per doctors instructions. When she did return to work she was on light duty for months.

Accident investigation findings

The day after the incident occurred the Human Resources Director investigated the accident. Investigation findings were:

- Kathy M. should never have been allowed to operate the feed lever. It was difficult enough for her to pull the feed levers that were installed correctly, let alone one that was incorrectly installed.
- Ms. M was wearing work boots that were 2 sizes too big for her. At the time of her initial employment the farm did not have the size she required.
- This was the first time that Ms. M. operated the feed levers. As she stated in a followup interview "the first and last."
- Kathy was never trained for this task and never told of the risks involved when performing the feeding process.
- The feed lever was incorrectly installed five years ago when the farm was originally built.

Figure 2.1-1: An employee, about the same height as the injured employee mentioned in the case study, attempts to grab and operate the lever.

Figure 2.1-2: Close up view of the incorrectly installed feed lever system. Notice the angle and position of the feed levers and how close they were to the ceiling. Compare to a properly installed feed lever pictured in Figure 2.1-3

Figure 2.1-3: A view of a correctly installed feed lever. Notice that an
extension handle could be used to reduce the overhead extension
which would be required for smaller employees.

RECOMMENDATIONS

Correctly installing the feed levers was a "quick fix." It did improve the situation, but there were still concerns about the whole process. The following are recommendations to correct the feed lever system:

- The best solution for this problem is to install an electronic/timed feed system. This system is used in the newly constructed farms at many similar facilities. This system automatically drops feed into the troughs at predetermined times during the day and requires no physical effort by the employees.
- Incorporate a portable lever extension piece into the feed lever system. This piece would allow the farm employee to operate the feed lever with less force. These extension pieces are used at other farms and processing plants.
- Conduct an inspection of all the farms that use the feed lever systems and make sure that the feed lever systems are installed correctly.

REFERENCES

Davis, W. T., Grubbs, J. R., and Nelson, S. M. (1995). *"Safety Made Easy: A Checklist Approach to OSHA Compliance."* Rockville, MD: Government Institutes, Inc.

Jacobs, P., Larson, N., and MacLeod, D. (1990). *"The Ergonomic Manual."* Minneapolis: ErgoTech, Inc.

Karp, G. (1996). "Ergonomic Job Design." *Onsight Technology Education Services* [On-line]. Available: http:// www.sirrius.com/~g...ht /articles/ergonom.html

OSHA. *"A guide to ergonomics."* (1995). *Division of Occupational Safety and Health.* Raleigh, NC.

CASE STUDY QUESTIONS

1. What are the three areas of importance in workplace ergonomic design?

2. What were the ergonomic hazards in this case study?

3. Identify the most important ergonomic hazard that directly resulted in Kathy's injury.

4. What were the ergonomic solutions recommended for the elimination of the ergonomic hazards identified in this case study?

5. List additional ergonomic strategies that might be implemented at this facility.

OVEREXERTION IN THE HOG INDUSTRY

William C. Hilton

INTRODUCTION

Back injuries alone cost the American industry 10 to 14 billion dollars in workers' compensation costs and about 100 million work days annually. Add the costs of hernias, sprains and strains to other body parts and it is evident that these types of injuries are the major accident related financial loss in the workplace (Ergoweb, 1996). Workers' compensation for these serious accidents resulting in sprains and strains range from an average of $6,899 in livestock production occupations to $3,647 in horticultural service occupations. Add to these figures the pain, suffering, and disruption to families and business and it becomes apparent that prevention of these problems must be addressed. Statistics indicate that this is a serious problem. Take for example the agriculture industry in Florida. There are over of 3,000 serious injuries to agricultural workers annually. (**Note**: A serious injury is defined as one which results in the worker missing one week or more of employment.) Approximately 75 percent of these injuries resulted from improper lifting, pulling, pushing (overextension) or from slips, trips and falls. Of these injuries, over 40 percent are sprains and strains; almost 50 percent of these injuries involve the back.

Table 2.12-1 presents the percentage of serious injuries caused by lifting, pushing, and pulling (overexertion) in seven major agricultural areas. These range from a low of 43.5 percent in horticultural production to a high of 65.0 percent in livestock production operations. The vast majority of these serious accidents results in sprains and strains.

CASE STUDY

On Friday, March 15, at approximately 1:00 p.m., Walter O., a 47-year-old male, had just returned from lunch. He was impatiently waiting for five o'clock to arrive. Walter had the weekend off and it was his first vacation in quite a while. As he surveyed the finishing house (the building where a hog

Occupational Area	Lifts, Pulls, Pushes	% of Accidents Slips, Trips, Falls	Other	Average Cost of Accident ($)
Livestock Production	65.0	17.0	18.0	6,274.00
Fruit & Vegetable	47.0	28.0	28.0	3,771.00
General Farm	51.9	30.4	17.7	5,372.00
Horticulture	43.5	32.1	24.4	3,647.00
Agri. Services	51.0	23.0	26.0	5,340.00
Livestock Services	60.0	26.0	14.0	6,034.00
Horticultural Services	51.0	12.0	37.0	6,899.00

Table 2.1-1: The percentage of serious injuries caused by lifting, pushing, and pulling (overexertion) in the seven major agricultural areas (Florida Agricultural Information Retrieval System, 1996).

is fed to full maturation), he realized that a 350-pound hog had died while he and his fellow co-workers were at lunch.

Walter, as he described himself, "has never been one to just sit around while there was work to be done." So he attempted to drag the sow out by her front feet. As he began to drag her out of her gate, Walter felt, what he described as, a terrible pull. He stood up and as the pain worsened he realized that he needed some help getting the dead hog out of its crate. Shortly after the incident his co-workers returned from lunch and removed the sow from the finishing house. When asked if Walter was okay his reply was "yea, I just stretched my back a bit." Walter walked gingerly for the rest of the day, until he bent over to pick up a piglet. When he bent over he said "it felt like somebody stuck a screwdriver in my back." When Walter felt the pain he knew he was in trouble.

The farm manager arranged for a co-worker to drive Walter to an urgent care center. The physician on call at the center diagnosed him as having a pulled muscle. The physician gave him a muscle relaxant and sent him on his way. That weekend Walter stayed at home, rested and applied a heating pad to his back. On Monday Walter went to work and proclaimed that he was "okay." That was until he bent over to put on his rubber work boots. This time Walter knew it was serious. He was rushed back to the urgent care center, where the same physician, again diagnosed him as having a pulled muscle. The physician gave Walter a muscle relaxant, ibuprofen, and a steroid dose pack. He also told Walter to stay off his feet, and to return for a follow-up appointment in one week. One week later Walter went back to the urgent care center; he showed no improvement. Since Walter was still in a lot of pain the physician sent him to an orthopedic clinic. Walter described his pain as starting in the region of his lower back and continuing to his groin and lower abdominal area. After X-rays and a thorough examination the doctor diagnosed Walter as having a herniated disk. Walter was prescribed physical therapy three times a week for four weeks, after which he would be re-examined.

After three weeks of physical therapy, Walter showed no improvement. A decision was made at that time to refer Walter to a neurosurgeon for a second opinion. After a thorough re-examination, the neurosurgeon recommended that Walter undergo surgery.

At the time this case study was completed Walter's claim was not closed. A Travelers Insurance agent indicated that Walter's injury could cost the company approximately $25,000. At the time when this case study was written, Walter's medical bills totaled $15,500 and were climbing.

Findings in the Accident Investigation

From January 1 to December 31 of that same year, 17 percent of the injuries recorded at this commercial hog farm were strains/sprains. These injuries accounted for 55 percent of the worker's compensation costs.

The following details were found during this investigation:

- Walter never should have attempted to move a 350-pound sow by himself.
- The farm was not equipped with a "come-a-long" to wench the sow out of her gate. (See Figure 2.1-4)

- The cart-boy, a device used to lift and transport dead hogs/sows out of the finishing houses, was too short to safely handle the animals.
- The aisles in the finishing house were to narrow.
- There were no exits in the middle of the finishing houses, so a hog that died near the middle of the building had to be carried to one end of the house.

RECOMMENDATIONS

The following are recommendations to correct the problem of moving dead hogs/sows from the houses:

- Future houses should have exits in the middle portion of the buildings. This would make it easier to remove the dead hogs/sows.
- Widen the aisles in the buildings. This will eliminate the use of a "come-a-long" and allow the employees to carry the dead hogs/sows out of the houses on the cart boys.
- Approach the engineer who designed the cart boy and discuss the needs of the farm employee(s). The carts must be wider and longer to safely handle hogs.
- Further investigate how other hog production operations confront this problem. Consider installing a winch system at each door and winching the animal to the exit where a forklift or front-end loader can pick up the animal and dispose of it in the gelatin dumpster.

REFERENCES

1. Jacobs, P., Larson, N., and MacLeod, D. (1990). "*The Ergonomic Manual.*" Minneapolis: ErgoTech, Inc.

2. ErgoWeb.(1996). "Manual Material Handling." *ErgoWeb*[On-line]. Available: http://ergoweb.mech.utah...td/fjw.html#Introduction.

3. FAIRS.(1996). "Accidents in Agricultural." *Florida Agricultural Information Retrieval System* [On-line]. Available: http://hammock.ifas.ufl.edu/txt/fairs/27221.

4. OSHA. "*A guide to ergonomics.*" (1995) *Division of Occupational Safety and Health.* Raleigh, NC.

Figure 2.1-4: A picture of the cart-boy, a device used to lift and transport
dead hogs/sows out of the finishing house. This device was too short
to safely handle the animals.

CASE STUDY QUESTIONS

1. How much is the average cost of accidents resulting in sprains and strains?

2. List the three leading occupational areas in agriculture for lifts, pulls and pushing accidents. What percentage of accidents are lifts, pulls and pushes for these industries?

3. Why did Walter get injured?

4. What was the nature and extent of these injuries?

5. What did the author recommend to eliminate the ergonomic hazards that caused this injury?

6. What would you recommend to eliminate the ergonomic hazards?

CHAPTER 2.2

CHEMICAL INDUSTRY CASE STUDIES

The chemical industry has many inherent health risks associated with the chemical raw materials, intermediate compounds, and finished products that are part of the manufacturing process. As a consequence, it would be very easy to overlook the ergonomically related health risks that are frequently present in those environments. The professional responsible for health and safety issues at a chemical facility should focus on the entire manufacturing process to ensure that ergonomic related problems are not ignored.

The case studies presented in this section could be found in a wide variety of industries besides chemical manufacturing plants. Research laboratories in both the academic and industrial environment could experience the same ergonomic illnesses reported in the following examples. If you have research or quality control laboratories, determine how the examples presented are similar to your facilities. What other strategies could be applied to modify the severity or eliminate the presence of the following problems?

ANTHROPOMETRICS IN THE CHEMICAL INDUSTRY

James P. Kohn

INTRODUCTION

This case study focused on body postures for obtaining samples of chemicals for quality assurance testing. The study was conducted at a large midwestern chemical manufacturing plant. The plant employed 4,500 individuals and manufactured 75 products. Many of the chemicals used in the manufacturing process at this facility could be classified as: carcinogenic, cytotoxic, mutagenic, teratogenic, corrosive, reactive, irritant, or toxic.

Raw materials arrived at the facility in loose bulk railroad cars as well as in fiberboard, polycarbonate or metal drums ranging from 18 to 38 inches in height. Up to 6 drums were stored on 4-inch tall pallets in a large warehouse. Forklifts were used to remove pallets from storage shelves in the warehouse and transport them to the appropriate locations in the plant. The drums normally remained on the pallets during sampling activities in the quality control area. Many of these drums had had chemicals removed and were less than half full.

STATEMENT OF THE PROBLEM

In order to reach the chemicals, employees in the quality control department were required to bend from the waist and put their entire arm or upper torso into the drum to scoop out samples. There was the possibility of back injury from bending or shoulder strain from reaching into the bottom of the drum to acquire a sample.

CASE STUDY

Participants

There were five quality assurance sampling personnel who sampled incoming raw materials and raw materials that had been previously opened and used. Three womsn and two men worked in this area. The heights of the subjects while wearing safety shoes were 5', 5'3", and 5'4" for the female employees and 5'9" and 5'11' for the male employees. All five employees considered their jobs to be potentially hazardous and were enthusiastic to learn of new methods for reducing the probability of injury or exposure to chemicals.

Procedure

A task analysis was conducted to evaluate the hazards associated with sampling raw materials from the drums. The shortest female was observed having to place her entire upper body inside a metal drum while obtaining a sample. The other employees were observed having to place part of their upper torso and entire arm inside several drums to reach the product.

Sampling task frequency ranged from twice per week to daily. A maximum of sixty samples were taken in one day. The largest sample size was less than six ounces; the weight and force required to obtain the sample was not a significant factor. A neutral body posture or working within a "safe operating window" is the preferred anthropometric position to minimize overexertion injuries (Grandjean, 1988). It is very important to avoid bent or twisted positions if ergonomically related injuries are to be eliminated.

To find the safe operating window for this sampling task, a series of steps were used to determine the best work posture as well as the best sampling tool design (Sanders and McCormick, 1993). The first step was to establish the safe operating window by determining the body dimensions important in the design of the sampling tool. The anthropometric measurements of elbow height and shoulder height were determined to be important for this project.

The second step was to define the population to use the equipment. The population consisted of the five employees working in the quality control department.

The third consideration was to design the equipment for the identified individuals in the sampling area (the stature range as well as the other key anthropometric measurements for these employees provided the adjustability range required for optimal tool design).

The fourth consideration was to determine what percentage of the population to accommodate. To accommodate 95% of all people in a 50/50 male-female workplace, the body dimensions of the 5th percentile female to the 95th percentile male were obtained. The fifth consideration was the type of clothing that employees were wearing. The employees, in this case, were wearing safety shoes which added almost an inch to their normal heights. This is important because most anthropometric tables provide data for individuals not wearing shoes. The sixth step was to find anthropometric data for the given population. The last step was to make a prototype of a new tool and/or redesign the facility layout and test it to determine if the modification would correct the problem.

RECOMMENDATIONS

Three options were considered to correct this problem: (1) mechanically tilting the drums, (2) purchasing longer scoops and, (3) designing a sampling tool. Option one was not considered acceptable because it would not eliminate exposure to the chemicals. Option two was not perceived to be practical since part of the sample taken would fall out of the scoop due to the angle required to remove the tool from the drum. It was decided that a sampling tool should be designed that could grab the sample required while, at the same time, eliminating the need for bending. All employees working in the area fell between the 5th percentile female height and 95th percentile male height according to standardized anthropometric tables. The acceptable working range for the population at risk (5th percentile female) would be no less than an elbow height of 36.9 inches and no more than a shoulder height of 47.7 inches. Employees should work with their elbows close to their bodies, within elbow height range, and no higher than shoulder height (Sanders and McCormick, 1993).

With these anthropometric measurements in mind, a sampling tool should have an adjustable design with a minimum length of no less than 11 inches and no longer than 47.7 inches. The sampling tool should be designed to have a handle grasp of no greater than 1.6 inches (Chaffin and Anderson, 1991). The collection portion of the tool should fit inside the smallest size sample container as a means of minimizing contamination to the container.

The Safety Department at this chemical plant is working with the engineering department to design a sampling tool. A model is under development and prototypes will be tested before implementing the new sampling tool.

REFERENCES

Chaffin, D. B., & Andersson, G. B., (1991). *"Hand Tool Design Guidelines." Occupational Biomechanics* (2nd ed.), (411-430). New York: John Wiley & Sons, Inc.

Eastman Kodak Company. (1983). *"Ergonomic Design For People At Work."* New York: Van Nostrand Reinhold.

Grandjean E. (1988). *"Fitting The Task To The Man."* (4th ed.) London: Taylor and Francis.

Sanders, M. E., & McCormick, E. J. (1993). *"Human Factors In Engineering And Design."* (7th ed.) New York: McGraw-Hill.

CASE STUDY QUESTIONS

1. Where did this case study take place?

2. What is a task analysis? How was it used in this case study?

3. What motions should be eliminated if overexertion-related injuries are to be avoided?

4. What were the steps used to analyze the ergonomic problem in this case study?

5. Which anthropometric measurements were important in this case study?

6. What anthropometric range of measurements was studied? Do you think that these measurements were appropriate? Why?

HAND INJURIES IN A CHEMICAL INDUSTRY QUALITY CONTROL LABORATORY

James P. Kohn

INTRODUCTION

This case study examined several employees who experienced hand pain. This problem occurred at a midwestern chemical manufacturing plant in the laboratory of the quality control department. This facility employed over 3,000 employees including 1,275 management and research staff. The employees of concern in this study reported the onset of hand and wrist pain after a large number of samples had been drawn and tested. An ergonomic evaluation was conducted including a task analysis and the leading contributing factor was determined to be the position of the hand during sampling and the frequent use of a pipette bulb. An automatic pipettor and alternating the use of each hand was recommended to eliminate this problem. A trigger finger pipettor was used for a short time, but it resulted in more pain in the employees than the manual pipette so its use was discontinued. This apparently was due tothe design of the automatic pipettor which required the depression and constant gripping pressure of the "pointer" or "trigger" finger when drawing or delivering a sample aliquot. Rogers (1983) reports that when the fingers apply force to activate switches, pressure increases in the skin and joint area of the hand. The employees took a varying amount of time off of work to recover from their injuries.

STATEMENT OF THE PROBLEM

The purpose of this case study was to examine the causes of hand repetitive motion injuries that resulted from pipetting, static loading of hands while filling test tubes, and finger grip tightening of clamps and caps during solvent extractions.

CASE STUDY

Participants

There were six employees that worked in the quality assurance laboratory at this chemical plant. Of primary interest in this case study was several of the female laboratory technicians who reported the ergonomic injuries. All the employees in this area had the same duties. Past medical records for these employees revealed that half had complained of shoulder or hand pain. The female employees of concern in this study had off-the-job hobbies including cross-stitch, knitting, gourmet cooking, and painting that contributed to their pain. They had curbed these activities as a result of the discomfort because they had difficulty holding the tools and utensils associated with these hobbies.

Procedure

An ergonomic evaluation was conducted in the quality control laboratory using a videotape recorder and 35 mm camera to record tasks performed by the employees in that area. Evaluation of the tapes and photographs, as well as interviews with the employees, revealed that pain occurred while tightening clamps that held the separatory funnels. Hand discomfort also occurred while tightening and removing the caps to the funnels as well as sample jars. Aliquot samples drawn during each shift ranged from 10 to 35 samples daily. A typical number of samples drawn was between 15 and 22.

Based upon an analysis of the videotape and the photographs, it was determined that the tasks associated with the jobs in this area required extensive use of the employees' hands. A task analysis was then conducted.

The maximum number of samples drawn took approximately two and a half hours to extract. A pinch grip using the thumb and the forefinger was used to grab and turn the clamp as well as the funnel and sample vial caps. The workstation was at the desired height for employees to be able to see the separation of the solutions and obtain the extracts while minimizing the frequency of bending. However, this workstation required employees to tighten the clamp and the cap at shoulder height. Analysis revealed that the wrist was twisted at an awkward angle during this activity due to the height of the equipment. Chaffin and Anderson (1991) documented how the angle of the wrist during gripping tasks affected the tendons and their synovia. The hand was found to be held with the fingers spread apart while pouring a solvent. This added to the stress placed on the employees' hands.

During the course of the workday over 180 minutes was spent pipetting using a bulb and 45 to 75 minutes using a thumb operator pipettor. When pipetting the hand was held in a static position. According to a study reported by Bjorksten, Almby and Jansson (1994), women who use plunger-type pipettors more than 300 hours per year have a fivefold greater chance of hand injury than women who, on average, pipetted less than 300 hours per year. The study was conducted in Sweden with 128 laboratory technicians of which 44 percent had hand and neck problems while 58 percent indicated that they experienced shoulder problems. The author recommended that a permissible exposure level of one to two hours of pipetting a day be set as a goal for laboratory analysis activities.

RECOMMENDATIONS

A handle grip should be used to grab the clamp and cap. An ideal grasp is about 1.6 inches (Chaffin and Anderson, 1991). Two of the six clamps on the shaker would have to be removed in order to add a grasping device unless the arms of the shaker could be extended. The angle of the shaker should be changed to eliminate the overhead positioning while twisting the cap. A narrow platform or step stool should be made available for employees to stand on during attachment of flasks and tightening of caps. A graduated cylinder, or smaller beaker, should be used when pouring the solvent.

Bulb, finger, and thumb control pipettors should be replaced with the better ergonomically designed automatic pipettors with a light touch activator switch. The quantity of solution needed could be dialed into the pipettor and a single light touch would activate the switch to pull or deliver the sample. This would eliminate static loading of the finger. A small beaker should be used for pouring instead of using a squirt bottle.

REFERENCES

Chaffin, D. B., & Andersson, G. B. (1991). *"Hand Tool Design Guidelines."* In *Occupational Biomechanics* (2nd ed.). New York: John Wiley & Sons, Inc.

Rogers, S. H . (1983). *Ergonomic Design For People At Work.* New York: Van Nostrand Reinhold.

Grandjean, E. (1988). *"Fitting The Task To The Man_."* (4th ed.) London: Taylor and Francis.

Sanders, M. E., & McCormick, E. J. (1993). *Human Factors In Engineering And Design.* (7th ed.) New York: McGraw-Hill.

Bjorksten, M. G., Almby B., & Jansson E. S. (1994). "Hand and shoulder ailments among laboratory technicians using modern plunger-operator pipettes." *Applied Ergonomics*, 25(2), 88-94.

CASE STUDY QUESTIONS

1. What physiological reactions occur when equipment switches are activated for prolonged periods of time?

2. What was the job classification of the employees involved in this case study?

3. How was the analysis of the job activities conducted?   List the advantages and disadvantages associated with the analysis used in this case study.

4. What physiological motions were identified as potential ergonomic hazards?

5. Identify the time and repetition factors that should be monitored to determine the possibility of elevated ergonomic health risks.

6. List the recommendations made by the author to eliminate the observed ergonomic risks. What would you recommend to control the ergonomic hazards identified in this case study?

CHAPTER 2.3

ELECTRONICS INDUSTRY CASE STUDIES

The material handling and repetitive motion problems presented in the electronics industry case studies that follow could have occurred in any industry. Any time employees are required to lift, pull or push objects, there is an increased risk of overexertion injuries. The NIOSH lifting guidelines examine factors presented in this case study. Pushing and pulling activities tend to be less frequently publicized in the literature, but they are just as much a factor in overexertion injuries as lifting motions. The factors that increase the probability of pushing and pulling related ergonomic hazards are also presented in this case study. What factors associated with these two material handling problems appears to be hazards that could be difficult to be identified? Repetitive motion is much the same as lifting. Performing repetition of motions while keeping the hands in awkward positions increases the risks of repetitive motion injuries.

The following case studies focus upon these two distinct types of ergonomic problems. Refer to an ergonomic textbook to obtain details on how to analyze these variables and determine the forces that increase the probability of ergonomic health risks. You might wish to select several tasks that are of concern to you and apply these analytical methods to those problems.

PUSH, PULL, AND LIFTING MATERIAL HANDLING ISSUES IN THE ELECTRONICS INDUSTRY

James P. Kohn

INTRODUCTION

This case study is based on a midwestern electronics manufacturing plant which employed approximately 2,100 people. Push, pull and lifting problems were identified in two locations of the plant: the electronics assembly area and the television packaging area. Injury frequency and severity rates were determined to be unacceptable by the safety staff.

STATEMENT OF THE PROBLEM

One of the problems studied at this facility was push and pull forces created while manually moving portable equipment. Portable equipment, such as carts or tool bins placed on casters, allows equipment to be transported that is too heavy to be lifted. Strains, sprains, slips, or falls can be incurred if the movement requires excessive push and pull forces.

The second problem examined the packaging area where lifting related injuries and muscles strains took place with excessive frequency. Four males, between the ages of 27 and 52, reported experiencing back pain and discomfort.

CASE STUDY

Assembly area

An employee was moving a steel-framed maintenance repair cart. The cart was mounted on four 3-inch diameter casters and weighed approximately 65 pounds. The employee was having difficulty pushing the cart and she resorted to pulling it. When the cart crossed an indentation in the floor, caused by a drain, the cart fell on her. The employee received several contusions; fortunately she was not seriously hurt. A dynamometer was used to measure the push and pull forces required to move the cart when it was empty and when it contained a full load of components. The average push force required was 51 pounds. The average pull force required was 42 pounds. The average force required to push was 62 pounds when full and the average force required to pull the cart was 73 pounds when full.

Television packaging area

A survey was conducted to determine lifting forces required to lift 72-pound large screen television sets from conveyors and place them into boxes for shipping. The NIOSH lifting guidelines were used to analyze this problem. The results indicated that employees working in this area had to handle excessively heavy loads.

Current information available in the literature considers lift, push or pull forces above 50 pounds to be excessive. The first example reached this limit and the second example exceeded the limit. The factors affecting the force required to push and pull objects include:

- Weight of the object
- Size and type of casters
- Frictional properties of the floor
- Stance at which the grip and movement is made
- Frictional properties of the soles of the shoes worn by the operator
- Body weight of the operator
- Leverage point for the operator to push off against

The factors affecting the lifting force required include:

- Weight of the object
- Size and handles (coupling)
- Origin of the lift
- Destination of the lift
- Distance traveled during lift
- Frequency of lifts
- Twisting motions during lift
- Duration and pace of the lifting task

RECOMMENDATIONS/SOLUTIONS

Assembly area

The same type of cart as those in the assembly area was used in another area of the plant. An employee familiar with them mentioned that the other area's carts were much easier to move. A dynamometer was used, once again, to measure the forces required for moving the cart in the other area. The average push force was 21 pounds and the average pull force was 19 pounds. The maintenance personnel in the assembly production area then had the same casters installed on their carts as were found on the carts in the other area. This greatly reduced the push and pull forces required to move the carts.

Television packaging area

Engineering was involved in developing a solution for movement of the televisions during the packaging process. A determination was made that current methods were unacceptable and the excessive force for movement could not be improved by modifying the conveyor system. Four vacuum lifters mounted on overhead tracks were installed in this area. The equipment had a suction area that attached to the top of the containers. Each unit had a capacity of up to 350 pounds. Short term results indicated that the company has been capable of maintaining its production goals in this area while reducing back injuries and muscle strains in the area by over 93 percent.

REFERENCES

1.   Chaffin, D.B., *"Workplace design to prevent occupational low back pain."*   Conference and lecture notes from the Occupational Ergonomics Conference, Denver, CO: August, 1993.

2.   Eastman Kodak Company. (1983). *"Ergonomic Design for People at Work,"* Volume Two. New York: Van Nostrand Reinhold.

3.   Mital, A., Nicholson, A. S., and Ayoub, M. M. (1993) *"A Guide to Manual Materials Handling, "* Bristol, PA: Taylor and Francis.

CASE STUDY QUESTIONS

1. Explain how push and pulling ergonomic risk factors are similar or different from risk factors associated with lifting.

2. What instrument can be used to determine pushing and pulling forces?

3. What is the recommended limits for push and pull forces? How much did the tasks involved in this case study exceed the recommended limits?

4. List and describe the factors affecting push and pull activities.

5. List and describe the factors affecting lifting activities.

6. Describe the procedures used to solve the ergonomic problems identified in this case study. What do you like or dislike about the solutions recommended in this case study?

7. What would you do to improve the methods used to solve this case study?

REPETITIVE MOTION PROBLEMS IN A SMALL ELECTRIC MOTOR MANUFACTURING FACILITY

Mark Friend

INTRODUCTION

An electrical component manufacturing facility located in the southern part of the midwestern United States, assembles and sells small electric motors for use in vacuum cleaners and other small electrical appliances. The company employed approximately 175 workers on a single shift. All 153 production workers were female and all earned minimum wage or slightly more. The first year of operations, the company had no discernible ergonomic problems. During the second year four cases of wrist pain were reported. In the third year the number of cases of reported wrist problems increased to 20, and to 53 by mid-May of the fourth year of operations. Two lawsuits were filed against the company, claiming that the employer was negligent in permitting working conditions that lead to abnormally high numbers of carpal tunnel syndrome.

Two consultants were called in to review the company's wrist problems and make recommendations. They learned that, as of May 17, the log showed 53 reported cases of wrist problems. The employer believed that many of the problems were exacerbated by off-the-job activities and that most of the complaining employees "viewed carpal tunnel syndrome as a ticket out of the plant with cash benefits."

STATEMENT OF THE PROBLEM

The problem of concern in this case study was the high frequency of repetitive motion induced carpal tunnel syndrome incidents. The problem facing the consultants was to determine if the employer's claims were true and to make recommendations to reduce the number of claims.

CASE STUDY

The consultants began by administering a questionnaire to all production employees of the facility. The survey requested information about such non job-related information as gender, age, height, weight, children and their ages, off-the-job recreational and avocational activities, and medical history.

All respondents were assured that any information given would aid in helping redesign the work place to better suit them and that each response would be confidential and would not be shared with the employer on an individual basis. Only cumulative data was to be reported to the employer.

Prior to administering the questionnaire, the consultants carefully researched contributing factors leading to the onset of carpal tunnel syndrome. Items were chosen for inclusion in the survey which could indicate predisposition toward developing wrist problems. All respondents were asked to give their names and even those who did not were identifiable through a process of elimination. A point system was assigned to questionnaires to weight them from high to low probability as to the consultants' perceived likelihood of the employees' developing carpal tunnel syndrome. What the consultants learned was that respondents who reported the highest predisposing factors were those who did not report any wrist problems. Those who reported few or no predisposing factors had reported wrist problems. The consultants concluded that workers who had reported injuries perceived that reporting contributing factors might lead to a denial of future benefits.

Following the administering of the questionnaire, the consultants did a walk-through and examined each workstation and most jobs in the manufacturing facility. Employees were found to be using small hand tools which were often inappropriately designed for the work. Poor postures resulted from poorly designed, sometimes makeshift, workstations. In one case, a worker was forced to reach more than four feet to grab one of the tools that she used during a repetitive part of her operation. Although she had submitted a request to the maintenance department to move a tool to make it easier to reach, she laughed sarcastically when asked when she thought the work order would be processed. Most employees expressed the belief that management and the people in the maintenance department were unsympathetic to their personal problems or those that pertained to their workstations. All seemed eager to vent their frustrations to the listening ears of the consultants.

RECOMMENDATIONS/SOLUTIONS

The consultants recommended that management implement a three-pronged approach to handle the problem. First of all, a complete investigation and redesign of workstations was implemented. New tools, chairs, and work benches were ordered and every effort was made to ensure that each worker had input regarding personal preferences. Secondly, a complete education program was implemented which began with training the management team and supervisors before workers were involved. Supervisors were instructed on causes of repetitive motion and similar types of ergonomic injuries. They were given the opportunity to learn steps that should be taken to avoid such problems. The supervisors were encouraged to give their employees the opportunity to rotate jobs and to follow up on employee complaints regarding specific workstations. Management insisted that the supervisors conduct the training for their respective work groups.

Thirdly, management was encouraged to engage in "case management" of each reported claim. Following a reported injury, the company nurse engaged in "handholding," to see that the worker received the attention that was needed for their satisfaction. The consultants stated that personal attention and transmission of a "we care" attitude would help alleviate many of the problems. As a result of the intervention techniques described, cases decreased dramatically. The state regulators used this plant as a model of what could be done to help solve carpal tunnel and similar ergonomic problems. Although the record of losses was not reduced to zero, management was satisfied that these steps provided an optimum return for their situation.

CASE STUDY QUESTIONS

1. Are symptoms related to ergonomic problems sometimes contagious? Is training the workforce about a problem like carpal tunnel syndrome likely to encourage or discourage the epidemic?

2. Was the consultant's approach appropriate for this situation? What other intervention techniques might have been useful?

3. How important is it to involve workers in the redesign process? Why does this always have to be considered?

4. In Sweden, carpal tunnel syndrome is not a compensible injury. As a result, there are hardly any reported cases. Is our system of compensation encouraging the reporting of exaggerated claims? What can be done to prevent this?

5. In this case study, management wanted minimal interaction between the consultants and the employees. All follow-up work was done through the supervisors. What do you suppose caused management to follow up this way and was this likely the best course of action?

6. Is there a psychology of ergonomics? If so, how was that utilized in this case and what could have been done to further improve the situation?

7. Was the survey of any value? Even if there had been a strong link between outside activities and reported claims, would it have made any difference to management?

8. Since the employees were covered by workers' compensation, did they have grounds for lawsuits? Is this a likely problem in other work places?

CHAPTER 2.4

HOSPITAL/HEALTH CARE INDUSTRY CASE STUDIES

The case studies that follow address the typical repetitive motion and overexertion ergonomic problems that could have occurred in any occupational environment. One of the case studies, however, addresses signal detection issues as they pertain to a fire alarm system. The signal detection case study is unique because it examines an alarm system that patients, staff, and visitors must recognize and respond to in the event of an emergency. While most industrial settings are primarily concerned with employees, hospitals have a larger population to be concerned about protecting in the event of a fire. Evacuation of non-ambulatory patients is an important issue for hospitals as well as geriatric facilities. Both of these settings face similar issues that are, indeed, ergonomic in nature.

After reviewing the case studies found in this section, consider the ergonomic problems identified. Study the recommendations and determine if other solutions could have been more effective at solving the specific problem. Determine if these problems are similar to those ergonomic problems of importance to you.

ELIMINATION OF BACK INJURIES IN A HOSPITAL
ENVIRONMENT

Andy Anderson, Barry Maxwell, and Eddie Johnson

INTRODUCTION

A county hospital served a 29-county area as a regional referral hospital and tertiary care center. The hospital has 725 beds and employs 4,200 people. Like many employers, this hospital was concerned about occupational ergonomics.

According to the National Institute for Occupational Safety and Health (NIOSH), citing the American Academy of Orthopedic Surgeons, the total cost to the country for back injuries is estimated to be between $20 billion and $50 billion per year. Back injuries are the second leading cause of absenteeism in the workplace. The Occupational Safety and Health Administration (OSHA) reports that 50 cents of each dollar paid as workers compensation applies towards back injuries.

In order to prevent these statistics from occurring at this hospital, an ergonomics subcommittee, which reports to the hospital safety committee, was formed. This committee was formed to identify and resolve ergonomic problems within the hospital. This project was initiated by members of the ergonomics subcommittee to assist in identifying and resolving reported cases of back injuries.

Under direction of the subcommittee, back injuries within the pathology area were identified as a major problem at the hospital. After reviewing the risk management department's incident data base, the special ergonomics team confirmed that this was an issue worth pursuing.

STATEMENT OF THE PROBLEM

Through a review of incident reports and work site analysis, performed by the physical therapy department, the back injuries reported were determined to be cumulative trauma disorders.

CASE STUDY

In addition to the physical therapy department's analysis, the ergonomics team conducted their own work site analysis. This analysis consisted of general firsthand observations and recommendations. It appeared that the pneumatic tube system, used for the transportation of specimens and information between areas, was contributing to the back injury problem at the hospital. The system has a network of 42 computerized tube stations which was installed in 1993.

The primary areas that used the system were: pathology, pharmacy, emergency, trauma, plant operations, admissions, and all nurse stations. Pathology was the department which had reported the most back problems due to a higher frequency of use of this equipment. The areas of pathology that used the pneumatic tube system were: hematology (1 station), urology (1 station), chemistry (1 station), blood bank (1 station), cytology (1 station), microbiology (1 station), and specimen acquisition (4 stations). During the team's work site analysis, they found that unloading the carriers from the pneumatic tube system was the major cause of back trauma.

All pneumatic tube stations were flush mounted in the wall with six-inch deep wells for receiving carriers. The bottom of this well was 22 inches above floor height. The padded carriers were four inches in diameter, 15.5 inches long, and weighed approximately 7/8 of a pound. This carrier was capable of carrying a 6.5-pound payload. Workers retrieved carriers approximately 30 to 40 times within an eight hour shift, three shifts per day. Once the worker retrieved the carrier, it was taken to a workstation where its contents were analyzed. Empty carriers were returned to the station and stored inside internal racks suspended within the unit.

The awkward posture required to retrieve a carrier, combined with the high number of repetitions, made this a job that would likely cause CTD. The specific problems were identified as:

- The well height was 22 inches from the floor, while the mean knuckle height for both males and females is 28.8 inches. This height difference requires the employee to bend. The team observed all employees using incorrect bending techniques.
- The flush mounting of the well combined with the low height of the control panel, requires extending the arm while twisting at the waste to avoid cranial contact with the control panel.
- The lower carrier storage rack, which is ten inches above the rim of the well, further hinders the access to the well.

- Employee's lack of knowledge in the proper methods of accessing the carriers.

RECOMMENDATIONS

The ergonomics team identified the height of the well and the accessibility of the carriers as the major problems. In addition, employee's lacked knowledge in the proper methods of accessing the carriers.

To raise the height, a tray was fabricated and mounted over the well projecting outward eight inches. This projection alone was sufficient to accommodate two carriers without reducing the original number which could be held. The lower carrier storage rack was removed and relocated. These modifications were implemented in the specimen acquisition area. Employees were instructed to stand directly in front of the station, bend at the waist, and retrieve the carrier with both hands.

The back problems were reported in two ways. One was voluntary attendance in the hospital sponsored back care class. The other was through a formal incident reporting process. In this case the worker and his supervisor filled out the necessary information on an incident report. This report was submitted to the employee health department and was evaluated by an occupational health nurse. Those employees who submitted incident reports were ask to attend the back care class. In both cases the physical therapy department evaluated the workers and the work areas. The physical therapist completed a job site analysis, which was forwarded to the employee, Ergonomics Committee, and was filed with employee health.

With approval from the hospital's ergonomics committee, the teams recommendations were implemented and are currently operational. The modifications were well received by both the ergonomics committee and all employees in the area. After a trial basis, these solutions will be implemented throughout the hospital.

REFERENCES

Eastman Kodak Company. *Ergonomic Design for People at Work*. New York: Van Nostrand Reinhold, p. 294, 1983.

"Ergonomic Improvements of Wire Wrapping Jobs: A Case For Injury Reduction," *Industrial Engineering*, pp. 48-50, January 1994, Roberta Carson.

R. Banham, "The New Risk In Ergonomic Solutions," *Risk Management*, pp. 22-30, May 1994.

CASE STUDY QUESTIONS

1. How extensive is the back injury problem in the United States?

2. What type of group was established to study ergonomic problems at this hospital? Are these groups common in the occupational environment? Should they be?

3. What type of analysis was used to determine the location and the extent of this ergonomic problem?

4. Where were the ergonomic hazards found to exist?

5. What were the key recommendations that were intended to eliminate the ergonomic problem?

6. What would you recommend to improve the redesign of the equipment?

ANALYSIS OF SIGNAL DETECTION ISSUES
IN A HOSPITAL ENVIRONMENT

Greg Baker and Rick Scott

INTRODUCTION

A signal can be defined as a stimulus or an occurrence that describes some aspect of the work process. The worker should be able to sense the stimulus, interpret the associated message, and respond effectively. Signals are usually auditory or visual in nature. The ability for an individual to sense and perceive a signal may mean the difference between death or survival.

Many times these decisions are made under conditions of great risk and are based on uncertain sensory information. The worker may also experience a decrease in monitoring and detection performance over the work shift. This may be due to the repetitiveness of a task or the number of times a person must detect a signal as part of their job responsibilities. There are several factors that will increase or decrease the probability that a signal will be detected.

Factors that will increase signal detection include:

1. Utilize simultaneous presentation of signals (audio and visual)
2. Provide two operators for monitoring
3. Provide 10 minutes of rest or rotate activity for every 30 minutes of monitoring.
4. Introduce artificial signals that must be answered.
   (These signals should be identical to real world instructions.)

Factors that will decrease signal detection include:

1. Excessive or insufficient signals requiring detection and response.
2. Introduction of a secondary display-monitoring task.
3. Introduction of an artificial signal when a response is not required.
4. Instructing the operator to only relay unambiguous signals.

CASE STUDY

This case study focused upon a fire alarm system that was located in a hospital facility. There were various signals associated with a fire alarm system. All of them were important for alerting staff, patients, and visitors if an incident did occur. In addition, fire alarm signals provide appropriate hazard response teams with important information if a fire were to take place.

This particular fire alarm system was relatively new. It was installed in 1989 as an upgrade for an old system. The cost of the upgrade was approximately two million dollars. The facility is an eight floor structure. There were six sirens on each floor. Each floor was divided into two sectors. There was a control panel on the ground floor in the security office. The major components of this system included the sirens and the control panel.

Sirens

The sirens were the source of the signal that most of the employees responded to during an emergency. Each siren emitted an 85 dB alarm when activated. Some of the sirens could be adjusted for lower sound levels for critical locations at the hospital. Each siren also had a flashing strobe light mounted on top of each wall unit. The system was adjusted to emit various types of "alarm modulations." This permitted the system to be set for different signals associated with fire, tornado or earthquake emergencies.

Control panel

The control panel was clearly marked indicating each sector of each floor. For example, sector "1" had a red and a yellow LED. When a smoke detector or pull activator in sector "1" was triggered, the red light illuminated. The yellow light was a signal for a general system malfunction. This could have been a fouled detector or an electrical problem. In the event that any message was sent to the control panel a siren sounded alerting those individuals monitoring the system. There was a small screen on the control panel that indicated which sector was sending the signal. A microphone, beside the control panel, was installed for announcements. The announcements were communicated through the siren speaker system.

General system operation

When one of the sensors or pull activators was triggered, the sirens were activated on that floor as well as the floors directly above and below. At the same time, a signal was sent to the control panel. Once the alarm was sounded security personnel had three minutes to reset the system or the alarm would then be sounded throughout the entire building. This gave security personnel time to dispatch someone to see if a false alarm had occurred. (They sent a security officer to the sector indicated on the control panel.) If it was determined that a false alarm had occurred, the system would be reset without alarming the entire building. At this point an announcement would be made stating that a false alarm had occurred. If it was not a false alarm, all sirens would be activated, and the building evacuated.

If maintenance or general repair work was performed on the system and it was inoperable, an announcement was made before and after the work was performed. The system was thoroughly tested twice a year.

RECOMMENDATIONS

Overall, this system appeared to be effective at alerting employees of a potential emergency. The number of sirens and noise levels were determined to be adequate. The strobe lights were also effective at alerting those individuals who may not be able to hear the siren. The control panel was well laid out and there was plenty of room for expanding the number of zones. The system would have been better if there was a way of determining exactly which detector or pull activator was triggered. This was an option that was available and was in the future plans when the budget would permit this purchase.

REFERENCE

Eastman Kodak Company. (1983) *Ergonomic Design for People at Work.*
    New York: Van Nostrand Reinhold.

CASE STUDY QUESTIONS

1. Define the term signal. Explain the role of signals in ergonomics.

2. What type of signals typically exist in the occupational environment?

3. Explain the process that workers go through when presented with signals in the workplace.

4. List the factors that will increase signal detection. How can a health and safety professional apply these factors when evaluating workplace signals?

5. List the factors that decrease signal detection. Why is an understanding of these deficiencies important in the study of workplace ergonomics?

6. What was the signal(s) of concern in this case study? Was it important to receive and correctly interpret these signals? Why?

7. Evaluate the quality of the system in this case study.

8. What would you recommend to improve this system?

9. Examine a signaling system. Determine whether the system of interest meets the criteria presented in this case study. If it does not, what can you do to improve the system?

MANUAL LIFTING CASE STUDIES
IN THE HEALTH CARE INDUSTRY

Barry A. Maxwell and Eddie Johnson

INTRODUCTION

A health care organization produced consumer health care products. This company has always had a philosophy of providing a safe and healthy work environment. The company routinely addressed ergonomic problems throughout its facilities by establishing programs that encouraged and supported employees to practice safer lifestyles on the job and at home.

STATEMENT OF PROBLEM NO. 1

Workers in the bandage-making department were exposed approximately six times each day to the risk of back strain. This was due to the requirement of lifting 54 pound rolls of bandages 20 inches horizontally in front of the body. Added to the weight of the roll was a 19 pound mandrel required to lift the roll. In addition, the worker was exposed to two severe pinch-points in loading the roll onto the unwind stand. The unwind stand base was built with a piece of angle iron running across the floor right where the worker needed to stand in order to insert the roll of bandages. This substantially increased the required horizontal distance in lifting.

   NIOSH Work Practices Guide for Manual Lifting verified that a problem existed in the loading of the bandage rolls . The workers had to lift a load that was too heavy for the distance required to hold the load away from the body. Machine operators had submitted suggestions for modifying this problem because they had experienced several close calls in straining their backs.

RECOMMENDATIONS FOR PROBLEM NO.1

The rewind stand and mandrel were redesigned so that the roll could be loaded onto the stand from the side rather than the end. This modification permanently attached the mandrel to the unwind stand reducing the amount of weight to be lifted by approximately 35 percent. Also, this allowed the operator to load the roll with the flat side against his or her body. This reduced the horizontal distance by half. In the new lift position the allowable weight limit was increased from 25 to 51 pounds, which resulted in a greater percentage of the workers being able to safely perform this job. The new design also eliminated finger pinch points.

STATEMENT OF PROBLEM NO. 2

Raw materials arrived at this plant in 50 to 80 pound bales. Filler materials arrived in 50-pound bags. Workers had to unload these products by lifting them from floor level pallets and placing the bags onto a conveyor that was three feet above the floor. Approximately 25 100-bags and bales were manually lifted each day. Surprisingly, the company averaged only two or three lifting-related back injuries per year. Production downtime, however, occurred with frequent regularity.

RECOMMENDATION NO. 2

The company's engineers installed a vacuum lifting system. The self contained lifting system could   lift up to 125-pound containers. The pneumatic suction hoist system ran on two overhead tracks and operated much like a crane. An operator squeezed the control handle to regulate the amount of vacuum required to lift the load. To lower a load, the operator engaged the control handle which decreased the vacuum. In the event of power failure, a safety system lowered the load without dropping it. The two systems transported 70,000 to 100,000 pounds of materials per day. No back injuries have occurred since the systems were installed. The vacuum system helped the plant achieve zero lost time from accidents. That has helped decrease the company's insurance cost as well as saving the considerable expense of workers' compensation claims.

CASE STUDY QUESTIONS

1. List the ergonomic risk factors associated with the lifting problem in this study.

2. What analytical technique(s) was applied in this case study?

3. What options did the author have available to eliminate the ergonomic risks associated with example one of this case study?

4. How can you determine when an ergonomic problem exists for a palletized product? List several solutions that could be applied to these type of problems.

5. What type of ergonomic control method was used to solve the problems identified in example two? Is this the best approach or would other control methods be superior? Why?

6. What are some of the possible benefits that one would expect after implementing the example two solution? List these benefits and explain the side benefits that an organization might expect to observe.

PSYCHOSOCIAL CASE STUDY
OF A MEDICAL TRANSCRIPTION UNIT

Kathleen Miezio

INTRODUCTION

The use of computerized monitoring of individual workers has made the scheduling and budgeting of work easier and quicker. When the computer counts the output of a worker, it can be color, gender, and disability blind. The worker knows exactly what is expected when it comes to quantity and quality of output. Performance evaluation becomes fair and objective. The good performers can be urged to keep up the good work, while the poor performers can be directed to exact techniques for improving their performance. All computerized monitoring systems for clerical work include a feedback system that allows the worker to access information about their performance against a standard at any time. This feedback can be used to improve the worker's performance.

Electronic monitoring of clerical work has made it possible for supervisors and managers to use human resources within the organization to the maximum. When a manager knows exactly how much a given worker will produce, there will be little dead time because all workers will be occupied with work; therefore, the worker is not under utilized or work does not fall behind. With computerized monitoring, there are hundreds of historical records on the demand for clerical services and the services rendered, so calculation can be made on how much manpower is needed for a particular unit.

For any enterprise trying to cut out waste and keep productive employees in good jobs, electronic performance monitoring of clerical work seems to be a great idea. It is an even better idea if the enterprise is a small hospital trying to survive in a small community—the added costs to health care, including the administrative costs must be kept in check for the health care institution to survive. Clerical work and administrative services, until the widespread use of computers, have been an invisible cost center that has been difficult to quantify and control. Now with the widespread use of computers, these costs can be kept to a minimum because managers of medical transcription services now have a means of controlling production.

STATEMENT OF THE PROBLEM

A group of medical transcriptionists (typists who transcribes clinic notes, correspondence, and other medical records) were exhibiting an unusual proportion of poor health. Five out of the seven transcriptionists had either been treated for Tendinitis of the wrist, upper neck pain, lower back pain or acute anxiety attacks. This work group was based at a small community hospital in the Midwest.

CASE STUDY

An ergonomist was called in to the unit to conduct an office ergonomic assessment. The ergonomist was asked to make recommendations to help alleviate the distress among the medical transcriptionists. The original hypothesis for the problems in the unit centered around the lighting and the design of the office chairs and workstations.

Subjects

Seven female medical transcriptionists worked in this unit. They transcribed clinic notes and correspondence from dictaphone earphones into a word processing system. All the transcriptionists had two year degrees in medical transcription, were aware of medical terminology and spellings, and had experience ranging from 2 to 13 years, the average being 6 years of service. The workers ranged in age from 31 to 60 years old.

Work design (process)

The transcriptionists earned from $9.00 to $11.00 per hour, a wage that is substantially higher for work in other industries. The work consisted of keying in clinic notes and correspondence for approximately 37.5 hours per week and the workers could use flex time. Peripheral tasks such as the printing of documents, answering the phone, transferring files, etc., were done by lower paid workers with less training in medical transcription.

The industrial engineering department at the hospital had recently time studied the transcribers and set a standard character count of 6996 characters per hour. This standard was set with a traditional time study using a stopwatch and the result demanded an output of about 40 words per minute, a typing rate that is reasonable for an experienced typist to achieve. A feedback system was built into the software and the transcriptionists could access their character count at any time. The transcribers had to also maintain a 2% or less error rate. Random documents were checked by hand by their supervisor for errors. Those who could not meet the demands of quantity and quality were reprimanded or not kept on the job.

Methods

Observations of the office were made with the transcribers working in the office. A brief survey of demographics, and discomforts (using the 1991 OSHA Meatpacking Guideline survey) was distributed to all workers in the unit. All the subjects were interviewed individually for approximately 30 minutes with questions about job characters, job demands, organizational dimensions and employee evaluation. These questions determined the quality and quantity of their task variety, task clarity, quality of work, feedback from the computer monitoring system, work hours, work load, work standards, task requirements, supervision, their performance evaluation, and job satisfaction. The interviews were designed to determine the job stress that was unique to this job.

Results

New office furniture had been purchased in the past three years and it was of superior quality—it was fully adjustable. Every worker in the office was given a chair of high quality with a well designed lumbar support. One worker, who was very overweight was simply too big for her chair. Another chair that was appropriately proportioned for a larger person was ordered promptly by management. While the office furniture and arrangements were quite adequate, the workers did not fit the furniture because they were never trained to use the furniture or its adjustments. All of the workers were dealing with less than optimum workstations that needed adjustments. The air quality in the office was excellent as well as the lighting. There was no glare on any of the video display monitors.

The interviews showed that this group of workers was under a great deal of psychological and social stress due to the design of their job. When asked what the workers liked best about their jobs, the transcribers liked the pay, the flexibility of hours (flextime) and the fringe benefits from working at a hospital. When asked what they like least about their job, the transcribers cited being "chained to the computer all day." Actually, the transcribers were given a mid morning and mid afternoon ten minute break plus a thirty minute lunch hour. Most transcribers skipped the breaks to double check on the quality of their documents or to "catch up" when they were behind on their character count. This resulted in almost constant sitting for long periods of time, which contributed to static loads in the shoulders, neck and back.

While the supervisors tried to be encouraging and understanding, they were under constant pressure to keep transcription costs to a minimum and provide "seamless" service to the medical departments. As a result, they used the results of the monitoring almost exclusively in employee evaluations for reprimands and the allocating of raises. Workers were actually threatened by the monitoring system, but, ironically they were also grateful for the flextime it offered.

The workers also disliked the monotony of their jobs and the amount of concentration that was demanded from them to do the job. This lack of qualitative variety plus a great quantitative demand on attention is typical of computer mediated tasks. The high workload demands were also found in the fact that the work was never really finished—there was always more in the buffer to transcribe.

The electronic monitoring system was a very serious stressor for this group of workers. They felt that the system acted as a "whip" to make them work harder. The social interaction with co-workers was eliminated, not by physical partitions between the workers, but by the demands of the monitoring system. All the workers reported that they felt "watched" by the system and that they felt isolated, although there was another worker only five feet away. This sense of isolation may be one of the reasons why workers on this unit experienced anxiety.

RECOMMENDATIONS

The physical ergonomics of the office was only part of the problem confronting management and workers in this office. Aspects of the design of the job had to be addressed along with the office furniture.

The ergonomist arranged to have a one hour training session on office ergonomics for all the workers in the unit along with the supervisors. Everyone was taught how to adjust the furniture and why the adjustments were important. Workers were also taught about the importance of moving around, taking breaks and stretching. The training emphasized the dangers of sedentary work, static postures and repetition. After the training, the ergonomist checked every workstation and helped to customize the fit of the worker to the workstation.

As with any issue that deals with the design of work and management functions, the question of psychological and social stress in the office had to be addressed. Management and the supervisors talked of "empowerment" of workers, but that idea of participation had little to do with the day to day work that went on. One supervisor's idea of empowerment was to allow the transcribers to choose the weekly "motivational" poster that was prominently displayed at the entrance to this office. The issues ran deeper, and the whole question of how and why computer monitoring was in use had to be examined.

The work group suggested that since a certain amount of work had to be produced every day, that the group as a whole, not individual workers could be monitored. That way, when a worker had a bad day, the other members of the group could help her catch up. The software could simply give the group cumulative character count to individuals and the supervisors. Since the workforce was stable, the supervisor agreed to the group monitoring as long as the character count remained the same as when individuals were monitored. It was agreed that any new transcriber would be monitored for three months during the probationary period, and the new person would be given a six month probation period.

The problem of the sedentary and monotonous feature of the job had to be addressed. The individual who printed and distributed the documents would be trained to be a transcriber. In turn, every transcriber would work in the printing and distribution of documents for part of their day. This would get the transcriber up and moving about. To address the monotony of the transcriber's job, a supervisory activity was taken over by the transcribers themselves. The transcribers spent a small part of their day checking other transcribers output for errors and feeding back results to their co-workers. This freed up several hours per day of the supervisor (who averaged a 55-hour week) to trouble shoot and interface with other departments. Furthermore, workers were encouraged to socialize and help each other solve problems during the day.

The adjustment to this type of job design and the teamwork involved was not an easy and painless transition; none-the-less, the unit has had no medical problems since the interventions and the unit has actually increased production.

CASE STUDY QUESTIONS

1. What were the ergonomic hazards of concern in this case study?

2. How does this case study differ from the other ergonomic studies presented in this casebook?

3. What factor did electronic monitoring have upon the health and performance of the employees?

4. Was the furniture a factor affecting the ergonomic symptoms described in this case study? Why?

5. What stress factors contributed to the work patterns that developed in this case study?

6. What do you think were the most important recommendations made by the consultant to address the ergonomic issues of concern in this case study?

7. What would you do to eliminate the ergonomic symptoms presented in this case study?

CHAPTER 2.5

MANUFACTURING CASE STUDIES

The ergonomic case studies found in this section cover a wide variety of health hazards and solutions for their modification or elimination. Repetitive motion hazards associated with the back, hand, arm, and wrist are identified in some of these case studies. Engineering and administrative controls were implemented to address many of these problems. In several case studies personal protective equipment (PPE) was attempted, but, as in most of the literature, wasfound to be ineffective in solving the ergonomic problems of concern. In one of the case studies, PPE actually created other ergonomic problems not initially observed among the employees.

In addition to the typical solutions to ergonomic problems, one of the case studies implemented a behavioral approach as part of an ergonomics process management system. When viewed as a whole, the recommendations presented in the case studies in the manufacturing section can provide the reader with a holistic approach to ergonomic problem solving. After reviewing the following case studies, identify the problems of concern, list the strategies employed to address those problems, and develop a list of actions that you would take to improve upon the recommendations presented.

A BEHAVIORAL ENGINEERING APPROACH
TO ERGONOMIC PROBLEMS

James P. Kohn

INTRODUCTION

The safety professional is the member of organizational management responsible for the prevention of losses associated with employee safety and health. Through the implementation of various management and engineering principles, safety professionals anticipate, recognize, evaluate and control those hazards in the workplace that affect the health and well-being of employees (1). Over the past several decades safety professionals have taken great strides at eliminating unsafe conditions in the workplace. The use of personal protective equipment (PPE) and engineering controls have steadily improved the safety professional's ability to reduce injuries associated with these hazards. People-related hazards or unsafe acts, on the other hand, have posed a far more difficult challenge for the safety practitioner (2 and 3). This becomes especially apparent when examining the ever increasing number of ergonomic-related injuries and illnesses (4).

Ergonomics is a Greek word that means "the laws of work" (5). It refers to a multidisciplinary approach which recognizes the physiological and psychological capabilities and limitations of people in the workplace (6). For some in the safety profession, ergonomics and people-related factors may appear as a new and highly publicized issue. In specific instances this is true since several examples of Ergonomics problems have risen from new workplace technology. The increased use of computer terminals in the office environment as well as more productive manufacturing methodologies on the shop floor have increased the requirements for employees to perform repetitive tasks. These "people-related" ergonomic problems may well be a reflection of industry's improved technological capabilities. In other instances however, the age old use of human strength for material handling tasks has simply compounded the exertion types of injuries that have plagued industry for centuries.

Explanation for the recent interest in ergonomic injuries goes beyond technology. Another factor to consider is that historically many employees may not have perceived Ergonomic injuries and illnesses as occupationally

This case study was presented at the American Society of Safety Engineers' Safety Technology 2000 Conference held in Orlando, Florida, June 19, 1995.

related. Numerous reports have indicated that many workers who suffered Ergonomic injuries in the past assumed that the cause of their afflictions was the result of off-the-job factors. In other situations, ergonomic injuries and illnesses that were thought to be occupationally related problems were, in fact, the result of off-the-job activities or conditions. Ergonomic problems are typically cumulative in nature and rarely the result of one incident. Factors on and off the job often contribute to the ultimate problem.

No matter whether the injury/illness results from factors on- or off-the-job, the occupational safety professional can not ignore ergonomics. If ergonomic hazards exist in the workplace, there is increased likelihood of ergonomically related worker's compensation claims. There is a steadily expanding body of evidence supporting the importance of ergonomic job design in the occupational environment. Losses associated with ergonomic injuries and illnesses in the workplace makes it clear why it has become the occupational safety and health issue of the 1990s.

According to the National Safety Council's publication "Accident Facts," occupational injuries as a whole resulted in costs estimated to be in excess of $111 billion dollars in 1993 (7). Almost $36 billion or 32 percent of those total injury costs resulted from ergonomic problems (7). The losses that result from Ergonomic problems are compounded when factoring in days away from work. In 1993, approximately 28 percent of the total estimated days away from work, or 21,000,000 days, resulted from Ergonomic injuries associated with overexertion (7). An additional 2,925,000 days or 3.9 percent of the total days lost from work resulting from nonfatal occupational injuries were due to repetitive motion injuries. The Ergonomic problem becomes a major concern to the safety professional when considering that 280,000 new cases of repetitive motion trauma occurred last year (8). This represents about 62 percent of all new occupational illnesses and about 4 percent of the total number of work related injury and illness according to Bureau of Labor Statistics estimates (8).

It is difficult to argue against the need to address ergonomic hazards in the occupational environment in light of these dramatic statistics. This is especially true for the safety professional when considering the proposed Occupational Safety and Health Administration (OSHA) Ergonomic Standard (9). In the minds of many professionals it is not a question of "is ergonomics a problem?" Rather, it has become a question of "What can we do about eliminating ergonomic hazards in the workplace?"

Considering the success safety professionals have had addressing unsafe conditions, it is tempting to address Ergonomic problems the same way. However, the PPE approach, in the opinion of most professionals and by federal agencies, is doomed to failure. Personal protective equipment has no impact upon the workplace conditions that cause the problem. It would serve as a red flag to compliance officers looking for poor workplace layout and inadequate job design. The other potentially costly control is to engineer the ergonomic hazards out of work processes and design. While engineering controls are the methods of choice when eliminating occupational hazards, they have two significant drawbacks:

1. Small companies may not be able to make the financial investments required for extensive modification of equipment or work processes, and
2. The best designed equipment can be used inappropriately if employees do not see the importance and benefits of the new equipment or process.

Several large companies with office ergonomics problems supports the second point. Companies have purchased the best ergonomically designed office equipment only to observe employees using poor workstation layouts. In most of these instances, the employees just did not know how to adjust the workstations correctly or that it was even possible to adjust their chairs, keyboards or monitors.

It is important to consider a third strategy to control ergonomic hazards. Some authors refer to this approach as a team approach to solving ergonomics problems (10). Others refer to this approach as a total quality management strategy (11). The underlying process is considered by this author to be a behavioral engineering approach to ergonomic problems (12). It is based upon the ergonomic objective: "the improvement of human health, safety and performance through the application of sound people and workplace principles" (13). The intent is to establish the best way to perform each task through input from all individuals involved in the activity.

By taking a "whole-job" approach, all the resources and responsibilities associated with a task are clearly delineated. As a consequence, it might be necessary to purchase new tools and equipment or change job processes to eliminate ergonomic hazards inherent in certain tasks. Before any action would be taken to modify and improve a task, all individuals performing or supporting the task would be involved in the task improvement process. An important component of this procedure, however, is the reliance upon sound data. No matter whether safety professionals view this approach as a quality management strategy as promoted by Deming and his colleagues (13) or a behavioral engineering approach modeled after occupational behavior researchers (14), the method is substantially the same. There must be a systematic approach for correcting Ergonomic problems based upon sound data.

Employing the behavioral engineering approach requires that the organization must:

1. Identify whether the potential ergonomic problem exists.

2. Evaluate the extent of the problem.

3. Establish standard operating procedures that interested parties have an opportunity to contribute to in terms of development and "buy in."

4. Establish a system to manage the ergonomic process.

5. Train everyone in the basics of ergonomic principles as well as roles and responsibilities required in the established process.

6. Provide continuous feedback to inform people of how they are doing in the achievement of the desired goals.

It is important to point out that this approach is not a safety program, but rather a safety process (15). A lesson that has been learned from TQM is that programs come and go, with little permanence in the organization. With the emphasis upon process, the behavioral approach to ergonomic problems becomes a long-term objective with continuous "mid-course corrections" using a NASA analogy. The potential benefits of this approach is clear when looking at the literature. Deming's 14 points suggests the importance of data as well as team participation and training in the process (16). The behavioral science literature points to the importance of personal involvement, workplaces that encourage the satisfaction of personal needs as well as group goal setting and feedback of performance in the achievement of desired performance (17). By integrating key elements of all of these approaches the behavioral engineering approach would seem to be a logical option for addressing ergonomic hazards in the workplace. The purpose of this study was to apply the key elements of a behavioral engineering approach to ergonomic problems in a small manufacturing facility.

METHOD

Setting

This study was conducted at a small, family-owned midwestern manufacturing facility that employed a total of 75 people, with approximately one-forth of the employees being female, working in three plants. Company management consisted of three plant supervisors, one for each building, and a maintenance supervisor. All the supervisors reported directly to the president and owner of the company. There was no safety or personnel manager for this facility. Safety related issues were handled by the plant supervisors.

The author and team of three associates became involved in this project upon the request of the company's worker's compensation insurance carrier. It was reported that the company's experience modification factor was well over 1.6, resulting in the concern expressed by the insurance carrier.

Upon inspection of the worker's compensation claims and OSHA 200 logs for a three-year period prior to the year when this study was initiated, it became evident that ergonomic injuries were of concern. Almost 31 percent of all of the occupational injury cases reported were sprains, strains, tendonitis and carpal tunnel injuries/illnesses. Incident frequency and severity data was calculated by job classification, location, task performed at the time of incident, age, gender, and experience/training. During the initial visit to this facility, supervisory and labor personnel were interviewed to assist in pinpointing the locations where the greatest frequency of Ergonomic incidents had occurred. As part of the interview process a medical symptoms survey was also conducted. Based upon the results of these initial activities, two locations became the focus of this project. These two locations were identified as the clear lacquer paint area and the bundling area, locations involving finishing and packaging processes.

The next phase of this project involved the detailed evaluation of tasks and machinery in these two areas. Tasks performed in these two areas were videotaped and analyzed providing an in-depth evaluation of the associated Ergonomic hazards. For example, weight of stock lifted during manual material handling was determined. Then key measurements associated with these manual material handling activities was recording permitting the evaluation of the lifting related activities in the two areas using the National

Institute of Occupational Safety and Health (NIOSH) 1991 Lifting Guidelines (18). In addition, repetitive motions in these two areas of Plant One were categorized and frequency measurements recorded for tasks identified as requiring motions which could result in ergonomic problems such as carpal tunnel syndrome.

Based upon the preliminary audit, risk assessment and Ergonomic evaluation, a report was submitted to management. Included in this report were recommendations for the feasible elimination of identified ergonomic problems based upon a collaborative labor and management approach. It was determined that performance observation checklists would be developed and trained supervisory personnel would perform random monitoring to establish baseline ergonomic performance for both Plants One and Two. To ensure consistency between the two observers, interobserver reliability was checked at the end of the supervisory training session. Slides depicting tasks associated with the finishing and packaging departments were presented and the supervisors were asked to record the number of ergonomic safe acts observed. To compute interobserver reliability, the percentage agreement method was used. The number of agreements was divided by the number of agreements plus any disagreements and then multiplied by 100. An agreement was counted when both observers categorized a posture or position in a similar manner. The raters consistently agreed to a great degree. The reliability averaged 87% for the six slides presented to the supervisors.

Baseline analysis

Plant supervisors were then asked to randomly observe employee safe ergonomic task performance in the two areas in each plant. Observations were made three times per day at random times during the shift. Safe ergonomic acts were tallied and averaged for each day to establish a baseline frequency of performance. While baseline analysis was performed a plan was initiated to establish plant safety committees that would serve as the receiving source of information for the ergonomic process. The plant safety committees were composed of representatives of both labor and management with decisions prioritized and presented to the president for his evaluation and support. Data obtained during baseline monitoring was shared with the committee as well as the strategy to be used to improve ergonomic performance.

Intervention

A 45-minute ergonomic hazard recognition and safe task performance training program then was conducted initially for the committee members and supervisory personnel and then for the employees working in the two target locations. A total of 54 employees attended the training programs. Five programs were conducted with groups of 8 to 12 employees attending each session. Training sessions included safe lifting methods and repetitive motion injury prevention. Following a review of the basics associated with lifting and repetitive motion, and to ensure worker ability to identify ergonomic hazards in their workplace, 15 slides of in-plant workers performing a variety of tasks were presented. These slides depicted safe and unsafe ergonomic task activities. Employees were then asked to identify the "good" or "not-so-good" aspects presented in the slides.

Following the slide presentation participants were asked to make recommendations for the improvement of the work process at their location. In addition, they were asked to make a commitment to assist in the process for correcting and eliminating the unsafe acts and conditions that were contributing to the ergonomic hazards. All groups agreed to work with the consultants and the safety committee in a team effort to reduce ergonomic injuries as well as to identify unsafe acts and conditions that could be corrected.

Concurrent with the training activities, recommendations for the modification of work place layout, machinery and work processes was solicited. In addition, machinery/equipment manufacturer representatives were invited to "walk around" the facility, interview employees, and identify changes that could be engineered to eliminate the ergonomic hazards. Results of these and the other activities were presented to the plant safety committees for their awareness and input.

During the final phase of this project a long term ergonomics/behavioral intervention strategy was developed to effectively manage this process. The safety committee working with this author developed a list of objects that they wished to accomplish to meet the long-term goals. In addition, data monitoring activities was established to provide ergonomic process feedback to the committee.

RESULTS

Evaluation of ergonomic hazards by work area

The ergonomic hazards for the tasks performed in both the clear lacquer paint (finishing) and bundling (packaging) areas is presented in Table 2.5-1. Included in this table is the task that was analyzed, the ergonomic hazards observed and the actual weight lifted versus recommended weight limits (RWL) as calculated using the 1991 NIOSH lifting guidelines.

Evaluation of ergonomic training effectiveness

Evaluation of Ergonomic training effectiveness was performed using two methods:
1. In class responses by participants to the ergonomic hazards depicted in the work area related slides, and
2. A post training program quiz consisting of 20 true and false questions.

Consistently, throughout all of the training sessions, participants correctly identified the ergonomic issues depicted in the slides. When safe ergonomic task performance slides were presented, participants were able to correctly identify the "good" aspects of the slide. These same participants were also able to correctly identify the "not-so-good" aspects of the slide as well as to recommend changes that would correct the situation. Results of the post training quiz supported the results obtained during the in class evaluation. The average grade for the post training quiz was 84.55 with the scores ranging from a low of 65% to a high of 100%. The most important result of this training program was that the majority of the employees agreed to assist the safety committee on this project.

Safe performance observation data

Baseline and post-intervention observations of the percentage of safe ergonomic performance was analyzed. Employees were observed working ergonomically safe between 47 and 78 percent of the time during the baseline observations. During the post intervention phase of this study, employees were observed working ergonomically safe between 67 and 92 percent of the time. This represented an improvement of safe ergonomic performance of approximately 26 percent. Performance charts were kept in the lunch room next to production charts for each specific area. Subjective responses made by employees and supervisors suggests that participants involved in this project were pleased with the improved performance. A good sign of the support for this project was that the charts were never defaced and supervisors reported that employees reminded them when they forgot to record results.

RECOMMENDATIONS

Post-intervention measurements indicated a significant increase in the frequency of safe ergonomic acts in both areas of Plant One and Two. Plant One, the facility with the greater initial frequency of unsafe ergonomic hazards, showed a superior reduction in unsafe acts and corresponding increase in safe ergonomic performance. Subjective responses by plant management and employees indicated support of the ergonomic approach that was implemented at this facility. In addition, safety committees were established in each plant to continue the ergonomic process as well as other safety related concerns. Involvement was perceived to increase commitment by employees to support and work for the success of the ergonomic process at this facility. Results of the various intervention activities suggested that employees understood how to recognize ergonomic hazards associated with their jobs and were committed to help increase the frequency of safe performance. Roles and responsibilities had been made clear and with the performance charts posted in the lunch room, employees knew how well they were doing in the achievement of ergonomic goals.

CLEAR LACQUER PAINT (FINISHING) AREA		
TASK	ERGONOMIC HAZARDS	Actual/RWL (1991 NIOSH GUIDELINES)
Crate Opening: Pry bars were used to open shipping crate and remove unfinished product.	*Shoulder Abduction, Overexertion, Strain due to pry bar activity. *Lifting hazard. *Ulnar wrist deviation while band cutting	A = 75 pounds RWL = 58 pounds
Paint Fill: A five-gallon bucket of paint was lifted approximately 5 feet off ground to fill paint application unit.	*Lifting hazard *Ulnar and radial wrist deviation *Overextension and unbalanced postures	A = 52 pounds RWL = 9.9 pounds
BUNDLING (PACKAGING) AREA		
TASK	ERGONOMIC HAZARDS	Actual/RWL (1991 NIOSH GUIDELINES)
Stacking: Product was gathered, lifted and carried to a banding machine where it was banded into bundles and stacked on pallets.	*Lifting hazard *Shoulder abduction, forward bending and overextension while static loading upper extremities *Ulnar wrist deviation and Dorsi/Palmar Flexion *Prolonged standing on concrete floor.	A = 38 pounds RWL = 27 pounds
Packing/Boxing: Product is lifted from storage bins, transported to the workstation where it is placed in boxes. Box lids are closed, stapled and placed on pallets.	*Lifting hazard *Shoulder Abduction, forward bending and over extension while static loading upper extremities. *Ulnar wrist deviation and Dorsi/Palmar Flexion. *Prolonged standing on concrete floor.	A = 16 pounds RWL = 8.7 pounds

Table 2.5-1. Ergonomic hazards by task in the clear lacquer paint and bundling areas.

The success of the behavioral engineering approach to ergonomic hazards at this facility was encouraging. Several supplemental strategies could be considered to increase the likelihood of success. Facilities should consider integrating their ergonomic activities into a comprehensive safety and health program. In addition, while emphasis must be made on the long-term benefits of the process, short term quick success is also critical to ensure support for the program. People want to see change and the implementation of a few quick success projects can help generate the long term support necessary for the accomplishment of the primary goal—elimination of workplace Ergonomic hazards. In addition, a system should be developed for employees to communicate Ergonomic concerns and suggestions for improvement. A weakness observed in this project was that messages were being lost between the employees in the areas and the safety committee. Overcoming this problem will increase employee participation and commitment.

One lesson to keep in mind is to learn from the results of your preliminary activities. Some activities are going to be successful while others will not produce the desired results. Keeping the concept of process at the forefront of your activities will assist you in learning from all of your activities. Continue to do what is successful and stimulate input to determine better ways to perform work related tasks. The most important lesson to learn from Deming is that it is important to "drive out fear" and "break down barriers" (19). Getting employees focused upon the task of improving Ergonomics in the workplace will achieve the ergonomics objective of the improvement of human health, safety and performance.

REFERENCES

1.   *"Scope and functions of the Professional Safety Position,"* Des Plaines, IL: American Society of Safety Engineers.

2.   Hammer, Willie, *"Occupational Safety Management and Engineering,"* Englewood Cliffs, NJ: Prentice Hall, 1989.

3.   Heinrich, H.W., Petersen, D., and Roos, N., *"Industrial Accident Prevention,"* New York, NY: McGraw-Hill, 1980.

4.   Manuele, F. A., *"On the Practice of Safety,"* New York, NY: Van Nostrand Reinhold, 1993.

5.   Eastman Kodak Company, *"Ergonomic Design for People at Work,"* Volume 1, New York, NY: Van Nostrand Reinhold, 1983.

6.   Plog, Barbara A., *"Fundamentals of Industrial Hygiene,"* Chicago, IL: National Safety Council, 1988.

7.   *"Accident Facts,"* Itasca, IL: National Safety Council, 1994.

8.   Bureau of Labor Statistics, *"Occupational Injuries and Illnesses in the United States by Industry, 1991,"* Washington, D.C.: U.S. Government Printing Office, 1993.

9.   Thurman, M.T., Alexander, D.C., and Smith, L. A., *"OSHA's proposed ergonomics standard,"* Professional Safety, 1994: 18 - 23.

10.   MacLeod, Dan., *"The Ergonomics Edge,"* New York, NY: Van Nostrand Reinhold, 1995.

11.   Walton, Mary, *"The Deming Management Method,"* New York, NY: Putnam Publishing Company, 1986.

12.   Kohn, J. P., *"Behavioral Engineering through Safety Training: The B.E.S.T. Approach,"* Springfield, IL: C. C. Thomas, Publisher, 1988.

13.   Kohn, J.P., and Friend, M. A., *"Quality and Ergonomics: The Team Approach to the Occupational People Factor,"* Professional Safety, 1993: 39 - 42.

14.   Komaki, J., Barwick, K., and Scott, L., *"A behavioral approach to occupational safety,"* Professional Safety, 1979: 19 - 28.

15.   Friend, M. A., and Kohn, J. P., *"A behavioral approach to accident prevention,"* Occupational Hazards, 1992.

16.   Crosby, P. B., *"Quality is Free: The Art of Making Quality Certain,"* New York, NY: McGraw-Hill, 1979.

17.   Kohn, J. P., and Timmons, D. L., *"Applying Health and Safety Training Methods,"* Springfield, IL: C. C. Thomas, Publisher, 1988.

18.   Waters, T. R., Putz-Anderson, V., Garg, A., and Fine, L. J., *"The Revised National Institute for Occupational Safety & Health (NIOSH) Lifting Equation,"* Bristol, PA: Taylor & Francis, July, 1993.

19.   Reilly, Norman, *"Quality: What makes it Happen?"* New York, NY: Van Nostrand Reinhold, 1994.

CASE STUDY QUESTIONS

1. Define the term ergonomics.

2. What technological advances have contributed to ergonomic problems in the workplace?

3. What percentage of workplace injuries are ergonomic in nature? What percentage of injuries are overexertion related and what percentage are repetitive motion related?

4. Explain the concept associated with a behavioral engineering approach to ergonomic problems. What is the basic premise or philosophy behind this approach?

5. Identify and describe the six steps that make up the behavioral engineering approach to ergonomic problem solving.

6. Describe the difference between a program and a process. Why is it useful to view the ergonomic issues in an organization as a process?

7. Identify the types of analysis used in this case study and explain how these methods were applied in this case study.

8. What is interobserver reliability? How is it used?

9. Why are baseline measurements important? How can baseline measurements be of use to the individual in charge of establishing an ergonomic process in an organization?

10. What were the intervention techniques used in this case study? Were they successful? How do you know if they were successful?

11. Do you consider the intervention strategy of establishing an effective management system to oversee the ergonomic process important to long term ergonomic success? Why?

ERGONOMIC CASE STUDY OF AN AUTOMOBILE RADIATOR PARTS MANUFACTURING FACILITY

Darla F. Hinnant

INTRODUCTION

The case study was performed at a radiator parts manufacturing facility. This company manufactures automotive after-market chemicals, under-the-hood and dash hoses, rubber products, and highway traffic safety devices. The automotive after-market chemicals include those products that individuals put into their vehicles after they leave the dealership. Under-the-hood and dash hoses and rubber products include products such as fuel lines, transmission lines, ducts for the heating and air conditioning in a vehicle and various other products.

Julie S., an occupational health nurse, began working for this company in 1992. Prior to her arrival, numerous complaints were received identifying problems at the hose division. After Julie. began working, she also began receiving complaints. Julie began to notice that several ergonomically-related recordable injuries were appearing on the Occupational Safety and Health (OSHA) 200 Log. This OSHA 200 Log indicated that the incidents were primarily occurring to the hand. As a result of the OSHA recordable injuries and the numerous complaints she received, Julie decided to consult management and begin an ergonomic study.

CASE STUDY

Her ergonomic study was performed at the company's hose manufacturing facility. This facility employed 14 people. At this division, hoses are bought in bulk and are cut into various sizes for different products, such as refrigerators, washing machines, and automobile hoses. The sizes of the hoses range from one-quarter inch (1/4") to one and one-fourth inch (1 1/4"). Hoses are manufactured from different materials, such as rubber and vinyl. The majority of the hoses packaged at this facility were rubber. After cutting the bulk product into user lengths, the hoses are then coiled and placed into packages. The employees at this division performed this task (cutting and packaging the hoses) for eight hours a day. Their eight hour work day was broken up by two (2) 15 minute rest breaks, one in the morning and one in the afternoon, and a 30-minute lunch break.

Julie noted that of the 14 employees in this division, three-fourths of them had complained about hand problems associated with this job. After reviewing the Occupational Safety and Health (OSHA) 200 Log, Julie noted that the recordable incidents, resulting from hand injuries, caused the company to lose a considerable amount of employee work time as well as money. Two of the injuries resulted in surgery for carpal tunnel syndrome (CTS) and the other employee, also diagnosed with CTS, decided to just grin and bear the pain rather than have surgery. The two employees who had surgery have since left the company. The third employee who decided to grin and bear the pain of CTS was still working for the company at the time that this case study was written, but she has been reassigned to another job. In addition to these three incidents, some employees have been diagnosed with arthritis and a few other employees have been diagnosed with CTS. However, those employees who have been diagnosed with CTS are constantly observed to make sure that their CTS symptoms do not deteriorate.

Procedure

Julie began attending seminars, obtaining books, and researching ergonomics hand, wrist and arm problems, such as carpal tunnel syndrome, tenosynovitis and tennis elbow. In addition to researching ergonomic injuries, she began interviewing the employees working at the facility. Julie looked at the longevity reports of the employees in this division and also developed a health survey for the employees to complete. She also spoke to the plant supervisor in charge of this division, Mack M., concerning the job related tasks, the number of complaints being received, and other critical issues related to ergonomic hazards. Mr. M. was an industrial engineer who felt that with his past experiences, he and Julie could develop a solution for this particular job.

Options considered to solve the problem

Julie and Mack considered numerous alternatives to eliminate the stress of repetitively coiling and packaging hoses performed at this division. At first, they tried the old-fashioned home remedies, such as ibuprofen, anti-inflammatory drugs and ice. Next, they decided to avoid hiring new employees that could be susceptible to wrist problems and injuries. Julie and Mack then bought wrist supports for all of the employees working on this particular job task at the facility. However, they discovered that the wrist supports helped some people, but did not help others. Some of the employees who were wearing the wrist supports began experiencing elbow problems, because they could not bend their wrists.

After these options failed to eliminate the problems from repetitively coiling the hoses and packaging them into containers, Julie contacted a consultant concerning a program that had been publicized in the local newspaper. The consultant had developed the ergonomic program as a result of similar repetitive motion tasks at their facility. The program consisted of performing stretching exercises. Mack and Julie determined that the exercise program the consultant promoted was not exactly what they wanted to correct their problem.

Although very frustrated at this point into the ergonomics study, Julie and Mack decided to look at other alternatives. The first thing that these two individuals considered was automating the process. After obtaining various proposals from different companies, they noted that upper management would never consider this option because of the substantial costs involved in automating the process. At this point they developed a make-shift ergonomic program, which included numerous components. The first component included alternating the different types of hoses that the employees were working with throughout the day. Next, the employees were rotated throughout different tasks during the day. Finally, some stretching exercises were developed by Julie and Mack for the employees.

RECOMMENDATIONS/CONCLUSIONS

In summary, Julie, the company nurse, and Mack, the plant supervisor/industrial engineer, looked at the problem at the hose division in great detail. After talking with the 14 employees in this division, these two individuals noted that the complaints, regarding hand injuries and problems, were legitimate and a solution had to be developed to reduce the number of injuries at this facility. They looked at numerous alternatives to solve this problem, such as old-fashioned remedies, wrist supports, stretching programs, and automating the process. After an intense review, the company decided to rotate the size and type of hoses an employee worked with daily and also rotate employees throughout different jobs each day. At the present time, Julie keeps a close watch on the employees. The three recordable incidents referenced in this paper caused the company to lose considerable production time and money. Consequently, this case study is strongly supported by company management and is on-going at the present time.

CASE STUDY QUESTIONS

1. How did Julie, the occupational health nurse in this case study, first become aware that a possible ergonomic problem existed at this company? What did she start analyzing to determine the extent of this problem?

2. What tasks did the employees at this facility perform that could contribute to the ergonomic injuries/illnesses observed in this case study?

3. How serious were the ergonomic illnesses experienced by the employees in this organization?

4. What did Julie do to become more competent at addressing ergonomically related issues? What would you have done to gain ergonomic knowledge and skill?

5. What were some of the strategies considered by the management team in this case study to eliminate the ergonomic hazards and reduce the associated ergonomic risks?

6. Did any problems occur as a result of the implementation of some of the ergonomic control measures? Identify those problems. Is this a common reaction to the implementation of workplace solutions? Why?

REPETITIVE MOTION PROBLEMS
IN A SYNTHETIC FIBERS PLANT

David A. Lawhorn

INTRODUCTION

Tasks requiring workers to perform highly repetitive motions have been one of the biggest ergonomic problems that industry faces today. Jobs demanding elevated rates of repetition require more muscular effort and, consequently, more recovery time than less repetitive tasks. Repetitive motion causes health problems for tasks that would normally be considered safe (Spilling, 1995). Repetitive stress injury (RSI) produces strains in virtually every joint of the body. This broad definition of RSI has opened the workers' compensation gates to the point where 30 percent of all new claims are RSI related, with each claim costing an average of $18,268 (Gustke, 1995).

There are many strategies to eliminating repetitive motion problems. This paper presents the need for an ergonomics program and four strategies that seem to be effective at addressing ergonomic problems. These four strategies are:

1. Employee training and involvement
2. Worksite analysis
3. Furniture, equipment and tool changes
4. Regular employee exercise

The program

There is overwhelming agreement among ergonomists that before any strategy can be successfully implemented, there must be an effective program. The first and most important aspect of starting any ergonomics program is obtaining management support. Support is essential in order to provide funding for new equipment, to convey willingness to try new methods, and to gather support from employees at all levels. This support can be secured through an ergonomics management education program. Part of this program should be based upon company-specific statistics. Prepare for this education program by gathering statistics on prior repetitive motion injury rates including how many have occurred, on what jobs, and what they have cost.

Once management has committed to supporting ergonomics, the next step is the development of a written ergonomics program. Effective implementation requires a written program that outlines clear goals, strategies for accomplishing them, and communication vehicles that are understood by all employees. Management support for this strategy should also be in writing within the program.

Finally, before the ergonomic program can get started, it is essential that an individual be dedicated to making it work. This responsible individual should be a trained ergonomist. If budget is limited and hiring a full time ergonomist is not feasible, a consultant or part time ergonomist should be hired (Carson, 1993).

Employeer training and involvement

Development of a training program is the most critical element in creating an effective ergonomics program. There should be continual communication between management and workers. When training employees each of the following topics should be covered: controlling risk factors; methods of prevention; detecting early symptoms of RSIs; the importance of reporting symptoms early; and appropriate work practices (Hequet, 1995). Through effective communication managers can inform workers on the goals of the program and the strategies that will be used to obtain them. Management can use this communication as an opportunity to educate employees on the ergonomically correct way of performing a task and the adjustments that could be made to make that task people friendly. In addition, workers tend to give extremely valuable suggestions for improvements since they usually have the best insight into their jobs. Workers can also aid in discovering physically stressful tasks that have not yet caused a reportable injury, but have been causing discomfort.

Employee suggestions tend to be the best ideas and are often the simplest and least expensive. Some suggestions might be as simple as changing the flow of work, moving the location of a piece of machinery, or job rotation (Cooper, 1995). In a pilot program at Michigan State University it was discovered that training, education, and information programs did, in fact, help raise awareness and reduce the severity of RSIs. Results from this study suggested that when people seek assistance earlier, those cases required less intrusive treatment and were easier to remedy (Springer, 1994).

Worksite analysis

The purpose of worksite analysis is to identify activities that contribute to RSIs and determine which jobs and workstations are the source of the greatest problems. To effectively analyze each job, it is necessary to understand the basic steps required to perform each task (Cooper, 1995).

Directly observing or videotaping an employee in action is a useful way to evaluate how a task is actually performed. If videotaping, slow motion, freeze frame and frame advance functions can greatly enhance the evaluation process. In addition, a tape can be replayed repeatedly, without further distraction to the workforce. If taping, start with a full body shot of the employee, including the surface in which they are standing or sitting on. Follow this by zooming onto the area of greatest concern. Be sure to observe the worst-case situation. There are many technical questions that can be asked while evaluating a task, but the most important question is, "Is this task necessary?" Task elimination is the most efficient control method for reducing RSI injuries. It not only removes causes of RSIs, it reduces associated costs as well (Stankevich, 1994).

After the basic steps are identified, hazards may be outlined. The next step is to determine the appropriate solution for each hazard. This is accomplished by changing the method of performing the task or changing the physical condition that creates the hazard. Finally, the analysis should be reviewed by a person that did not participate in the observation. This review will generate questions that can be used for fine-tuning the ergonomics program (Cooper, 1995).

Furniture, equipment, and tool changes

Ergonomically designed tools and equipment that take into account both psychological and physical aspects increase job satisfaction and prevent injuries. Tools and equipment should minimize extreme awkward body postures and movements. Ergonomic factors that go into tool design include weight, shape, vibration, noise, grips, switches and pressure on the hand and arm posture.

The key word for reducing RSIs through furniture is adjustability. Furniture should make the user feel comfortable, and the user should be able to sit or stand in a variety of positions. An adjustable keyboard support, work surface, or chair are some examples of creating better support for computer workstations (Cooper, 1995). Many companies are starting to look at alternate computer input devices as a means of proactively managing RSIs. One of these devices is a pen-based system for people with severe RSIs. It permits data input all day long with no discomfort. The ultimate in input devices is voice recognition systems where a person's job can truly become a hands-off experience. Most individuals can use a voice recognition system for three hours straight without experiencing fatigue. The systems available today gradually "learns" the vocabulary of the user and becomes more effective over time (Hilton, 1994).

Regular employee exercise

In response to increased incidence of RSIs, interest in upper body exercise programs have flourished. These programs typically take place on the shop floor or in the office and involve several breaks from the work routine. Such programs are not considered corporate fitness or wellness programs, which usually take place outside of working hours and focus on general conditioning and cardiovascular fitness. The idea behind this type of program is that training yields stronger, less fatigued muscles, while providing more stable, less injury prone joint structures. Passive stretching is thought to return ligaments, tendons, and muscle contractile to a range where risk of injury is minimal.

An ergonomist who wishes to include an exercise program as part of a RSI intervention strategy should consider several factors. First employees should be screened for pre-existing medical conditions in order to avoid exercises that may be harmful to them. The exercise program should consider the particular work stress the employees are under and be tailored to problematic job tasks. At the same time the exercise should not duplicate problematic elements of the employees job such as static or repetitive contractions. The selected exercises should not be embarrassing to perform or disruptive to the workplace. It must be pointed out that exercises that involve stressful motions or reduce rest periods may actually be harmful to the employee (McGorry, 1995).

CASE STUDY

This case study was conducted at a synthetic fiber manufacturing plant. The polyester manufactured at this plant was used to make seat belts, clothes, sheets, and many other products. This plant produces polyester from semi-raw materials and sells it in two basic forms, polyester that is cut into small pieces and polyester that is spun as thread onto tubes as a continuous filament.

According to the plant safety director, the preparation of tubes that thread is spun onto involves one tube printing machine that is operated by two people in charge of printing product identification onto tubes. The machine is operated in three stages. (Refer to Figure 2.5-1 for a schematic of this operation.) First, one person loads the tubes by hand into a hopper; second, the tubes are printed with identification; third, a person receives the finished tubes and stacks them into a bin. (Refer to Figure 2.5-2 for photographs showing this process.) These two jobs are rotated evenly between the same two people throughout the shift. Currently, this job requires an average of 35 bins to be filled in an eight-hour shift with approximately

500 tubes in each bin. This requires each person to load the hopper with 8750 tubes per shift by reaching, grasping by hand, and twisting their body back and forth between the hopper and the pallet of tubes. Each person also has to stack the finished tubes into bins which requires an additional 1458 lifting and twisting motions between the tubes coming out of the printer and the bin.

This type of activity is likely to cause epicondylitis (tennis elbow), which is an inflammatory reaction of tissues in the lateral elbow region, and carpel tunnel syndrome (CTS), which is a wrist injury caused by compression of the median nerve inside the carpel tunnel within the wrist. Symptoms of CTS include tingling, pain, or numbness in the thumb, index, middle, and ring finger (Hequet, 1995; May and Schwoerer, 1994). Furthermore, one worker that has done this job for four years was diagnosed as having tennis elbow. There has also been numerous complaints of finger soreness in the thumb and first three fingers among workers in this area. Both of these disorders are categorized as RSIs and are considered preventable and treatable through ergonomic improvements (Vasilash, 1994; Powell 1995).

There has been three diagnosed cases of cumulative trauma disorders and numerous RSI symptoms within the area in which tubes are prepared. This study examined the alleviation of RSIs caused by preparing the tubes that thread is spun onto by implementing ergonomic improvements into the process.

RECOMMENDATIONS

Strategies used to eliminate RSIs

The strategies that have been implemented in this department included worksite analysis, employee involvement in the process, and minor tool changes. A checklist was used to pinpoint the ergonomic risk factors associated with upper extremity RSIs. Photographs were also taken of the human motions and extensions that were taking place. Employees were also asked to identify what procedures or tools they thought would alleviate pain. Two ideas generated from these interviews was to change the tools that were used and rotate employees more often. The following items were implemented in this department:

1. Rotate bin and hopper job among other workers in the area thereby reducing the time each employee is on each job.

2. Initiate the use of a two handled fork, utilized for loading bins, thereby distributing the product weight to two arms.

3. Distribute gripping gloves thereby reducing grasping forces on the hands when picking up tubes.

4. Survey employees informally that have suffered from ergonomic problems and identify what they have found to be effective at eliminating the causes.

The following recommendations were also considered for the elimination of ergonomic stressors in this work area:

1. A forklift truck could deliver the tubes directly into the printing machine if the hopper were modified.

2. A chute could be fabricated to feed the tubes directly into the load out bins.

CONCLUSION

It was discovered that job rotation was not widely accepted among the employees in this department. This was because workers did not like moving to another job that they were less familiar with, even if their job was more difficult. The two handled fork (tool modification) was embraced by the employee who came up with the idea; everyone else went back to the one handle. The gripping gloves were embraced by all and show signs of immediate tension relief. There was minor improvement among the employees that were suffering from tennis elbow, which was thought to be result of the use of the two handled fork.

Each of these strategies become exponentially more valuable and effective when used in conjunction with other strategies. Additionally, it is important to remember that any program must continually be updated as changes are made within the facility. More changes are expected in the future.

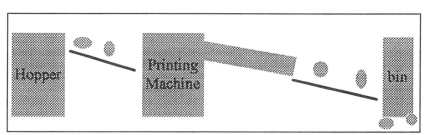

Figure 2.5-1: Diagram of the equipment found in the tube preparation area.

Figure 2.5-2: The top photograph shows the stacks of tubes as delivered to the tube printing area. The bottom photograph shows an employee grasping several tubes at a time to load the tube printing machine.

REFERENCES

Carson, R. (1993, September) "How to start a successful ergonomics program." *Occupational Hazards,* 122-127.

McGorry, R.W., & Courtney, T.K. (1995, June) "Exercise and cumulative trauma disorders." *Professional Safety,* 22-25.

Hequet, M. (1995, May) "Ergonomania." *Training,* 45-50.

May, D.R., & Schwoerer, C.E. (1994) "Employee health by design: Using employee involvement teams in ergonomic job redesign." *Personnel Psychology,* 861-876.

Vasilash, G.S. (1994, July) Ergonomics, "The blinding flash of the obvious." *Production,* 47-49.

Powell, M.T. (1995, March) "Here's how to keep injured workers productive." *Managing Office Technology,* 16-18.

Cooper, R. B. (1995, September 25) "Worker-workplace fit is key goal of ergonomics." *National Underwriter,* 9-19.

Gustke, C. (1995) "Corporate stress over repetitive strain." *Institutional Investor,* 31.

Springer, T.J. (1994, March) "Managing effective ergonomics." *Managing Office Technology,* 19-24.

Stankevich, B.A. (1994, May) "Guidelines for videotaping and evaluating CTDs. " *Professional Safety,* 37-40.

Hilton, D. (1994, August) "Alternate input devices take a load off wrists." *Managing Office Technology,* 32-33.

Spilling, E.S. (1995, August) "Work-related upper limb disorders." *Work Study,* 27-28.

CASE STUDY QUESTIONS

1. What are the four strategies presented in this case study that the author feels are effective at addressing ergonomic problems?

2. What are the key elements of an effective ergonomic program, according to this case study? Do you consider these elements important? Why?

3. In many case studies in this casebook, authors discuss the importance of retaining a trained ergonomist. Do you think that this is important? Why? How do you locate a trained ergonomist?

4. What role do employees play in an ergonomic program? What usually determines whether you will secure employee support?

5. Evaluation always plays an important role in determining program success? Is this true for ergonomic programs? Why?

6. What is the key word associated with ergonomically designed workstation furniture or equipment? Why is this concept important?

7. Is employee exercise programs important? What data is present in the literature to support the use of exercise programs for the control of ergonomic problems?

8. What was the ergonomic problem in this case study? What recommendations were suggested to eliminate these problems? What would you do to modify the equipment or work area to effectively control these problems?

REPETITIVE MOTION PROBLEMS IN HANDPACKING SYSTEMS

Leslie C. Ridings

INTRODUCTION:

Performing repetitive tasks, or tasks that involve the use of vibrating tools, without taking the proper precautions could result in problems that may range from mild daily pain to serious muscular and nerve damage. The pain associated with repetitive motion injuries (RMI) most often affects the tendons, nerves and muscles of the hands, wrists, elbows, and arms.

There are many different kinds of repetitive motion injuries. Three of the most common are known as trigger finger, carpal tunnel syndrome (CTS), and tennis elbow. All of these conditions can be painful and even debilitating.

A health care paper manufacturing plant noticed repetitive motion problems relating to their handpacking system. Repetitive motion of the hands, wrists, and back were causing problems for the temporary employees. The temporary employees were not paid by the plant. However, the company still worried about possible long-term problems associated with these pains.

The purpose of the handpacking system was to place personal hygiene pads into cardboard boxes then use a forward push onto a conveyor belt so the box could be taped and sent to the warehouse. One employee retrieved a cardboard box from the floor, loose folded the box, and placed it onto the roller conveyor. This same employee then retrieved eight to twelve packs of pads, depending on which line they were working on, and placed them into the cardboard box. The employee then pushed the box onto the conveyor belt. This belt took the box to a machine that taped the box and sent it to the warehouse.

The problem with this system was the repetitive motions performed by the hands, wrists, and back. The continuous twisting and bending performed when the employee retrieved the box from the floor caused low back pain. The shoulder and wrist pain were caused by the three following tasks performed by the employee:

- Loose folding the box.
- Placing the personal hygiene pads into the box.
- Pushing the box onto the conveyor belt to be taped.

The shifts for handpacking in this area of the plant lasted eight hours. The employees had two 15-minute breaks and one 30-minute lunch break. This meant that the employee handpacked for a total of seven hours. In these seven hours, the following repetitions are performed:

- Pinch Grip
- Finger Press
- Radial Deviation
- Ulnar Deviation
- Back Twists

CASE STUDY

The module safety manager, an occupational health nurse, and the plant's safety leader, were involved in the process of determining the hazards of this handpack system. The module safety manager videotaped the handpacking system with a camcorder and the entire team assessed the problem together by interviewing the employees, watching the task being performed on the production floor, and by watching and counting the repetitive motions being performed on the videotape.

In the article "Preventing Repetitive Motion Injuries" by Elizabeth Sheley, the importance of videotaping was emphasized. The article stated that "...videotaping employees at work is crucial. Videotapes allow safety consultants to analyze the problems thoroughly and can serve as useful training tools." This was true for this plant. Videotaping allowed the team to discuss possible techniques that would eliminate the repetitive motions being performed. It also allowed the team to get a baseline measure of how many movements were actually taking place.

The company wanted to look at all possible options where the handpacking task was concerned. The team researched the problem and found five steps to preventing carpal tunnel syndrome, a type of repetitive motion injury. The five steps listed below were found in the article *Dissecting the CTS Debate* by Susannah Zak Figura:

- Be sure workstations (tables, chairs, and equipment) are at the appropriate height to minimize musculoskeletal stress.
- Redesign jobs to reduce the amount of repetitive motion.
- Provide workers with opportunities for adequate rest periods.
- Keep production expectation realistic.
- Teach workers the importance of taking breaks and of using tools correctly.

These five items were looked at closely when trying to figure out some possible solutions for this problem.

RESULTS

The company's safety team had to find quick solutions to the handpacking problem. When all the results from the videotape were calculated an acceptable handpacking task frequency was formulated for the employees. From the calculations from this study as well as data given from other sister plants, it was determined that less than 7500 total movements in a shift period yielded the best result and kept complaints down. The following is the information that was found:

- A total of 4886 repetitive movements were made by the employee per shift.
- A total of 3728 hand-wrist movements were made by the employee per shift.

RECOMMENDATIONS

From this information, the company's safety team recommended the following:

- Rotation should occur after one hour or less of handpacking or repacking.
- The rotations should result in a person using a different motion and different posture. In other words, the ergonomic stressors should be shifted to another body part. If a person was using a pinch grip and flexing both wrists frequently, they should rotate to a job that required much less wrist stress and perhaps more elbow/shoulder repetition.
- Some type of slanted device, perhaps the one used at other sister plants, should be used in order to bring the boxes up to waist level in order to decrease bending and lifting.

The problem with these recommendations was that there were not enough employees to compensate for the amount of time spent handpacking, nor were there enough small jobs for the employee to do while the other person was handpacking. Further investigation has been underway in order to solve all the ergonomic problems in this area and find more effective solutions.

The problem of handpacking has not yet been resolved. More investigation is needed in order to figure out an adequate amount of time an employee can handpack on each line without problems and to try to develop the system where everything is waist high and ergonomically situated.

Overall, the health and safety team is testing several options that would help the handpack system reduce future repetitive motion problems. More emphasis needs to be placed on:

- *Appropriate heights of the workstations*—everything needs to be waist high so the body will be aligned.
- *Adequate rest periods for the employees in order to be able to rest different groups being used repetitively.* This can be done by rotating the job as mentioned earlier, or by taking breaks at regular intervals.
- *Keeping production expectation realistic.* Employers need to realize that certain jobs can only be performed at a certain pace to be healthy and safe for the employee. These repetitive motions being performed need to be done at a safe and slow enough pace so the employee will not have any long-term medical problems associated with repetitive motion injuries.
- *Help the employee redesign the way they are picking up boxes and packing the pads before a realistic goal can be set.* Holding ergonomic classes for these employees and teaching them alternative ways to perform these tasks would cut some of these repetitive motions they are performing drastically.

Repetitive motion injuries are a growing problem for industries today. According to the Bureau of Labor Statistics (BLS), in 1993 carpal tunnel syndrome causes 41,019 private industry workers to miss days on the job. Therefore, adequate solutions to the handpack problems that the temporary employees were experiencing are needed so the company does not contribute to the carpal tunnel statistics reported by the BLS each year.

REFERENCES

Figura, S. "Dissecting the CTS Debate." *Occupational Hazards*: 28-32, 1995.

Sheley, E. "Preventing Repetitive Motion Injuries." *HR Magazine*: 57-60, 1995.

CASE STUDY QUESTIONS

1. Who were the workers of interest in this case study? Why does this population pose an interesting problem for organizations in today's business climate?

2. Identify the tasks and the required motions that employees performed that contributed to the ergonomic problems observed in this case study.

3. List the job classifications for the individuals that served as the ergonomic committee in this case study. Was this a good representation of the important or key ergonomic stack holders at this company? Why?

4. List the benefits of using videotaping and photographic tools in an ergonomic program. What are some of the benefits or disadvantages of these tools that were not mentioned in the case study?

5. This case study refers to a five step process identified in an article written by Susannah Figura. List these steps and explain the activities associated with each of them.

6. What was of interest to you regarding the solutions recommended to solve the ergonomic problems in this case study? What role did frequency of motion measures play in this process? Would you have incorporated other measures in your recommendations? What would they have been?

A ERGONOMIC CASE STUDY
IN A SERVO-MOTOR ASSEMBLY FACILITY

James Coltrain and Barbara Dail

INTRODUCTION

The manufacturing facility examined in this case study employed 70 workers. The facility is located on 21 acres of land. The manufacturing building is 280,000 square feet, 16,200 square feet of which is office space. The manufacturing area in operation includes: an injection molding area, an electronic washing area, a production area, a machine shop, a welding area, a quality assurance lab, a vibration room, an office area, a hazardous material storage building, a warehouse space, and a minster press area. There was no formal ergonomic program in place at the time of this study. All ergonomic activity was handled in an informal manner.

STATEMENT OF PROBLEM

The purpose of this project was to identify ergonomic stressors that may cause repetitive motion injuries to operators of a servo motor assembly line. To pinpoint a specific job task, an informal interview was conducted with the assistant manager of a servo motor assembly line. The primary ergonomic problem was associated with contact plate assembly. Fine motor skills were found to be required to perform this task. Frequency of motion analysis indicated that five repetitions of motion were required per minute at each standing workstation studied.

CASE STUDY

Several sessions were spent at the assembly line observing the assembly process (job task analysis). During these sessions, the following stresses or contributing factors were noticed:

Stress to the body

Neck — Neck angles downward (work platform too low)

Back — Operators bending over (work platform too low)

Wrists — Wrists held at awkward angles (deviation from neutral position)

Arms — Static loading of arms (arms elevated above the work surface for extended periods of time)

Eyes — Eye strain (due to small size of parts)
Legs — Legs sore (due to standing operation)
Fingers and hands — Fatigue and loss of finger flexibility
(continuous holding of small parts)

Additional factors noted during the evaluation of the workplace were: worker complaints, quality problems and low productivity.

The company used the method of rotation of job assignments every 2.5 hours. It also designed and installed an assembly jig in an attempt to address the ergonomic problems.

Interviews were conducted with the operators on the assembly line to determine the effectiveness of the ergonomic solutions. Job rotation was seen as a very effective measure to relieve stresses on the neck, back, wrists, arms and eyes. The down side of job rotation was that the jobs the operators rotated to also required the operators to stand.

The assembly jig, which was designed to relieve ergonomic stresses, appeared to have created even more problems. The operators did not like to use the jig because it required the parts to first be picked up, aligned and then slid into place. This new process required several additional "awkward" wrist positions and added an extra step to the process which reduced production rates.

RECOMMENDATIONS

To relieve or eliminate the ergonomic stressors associated with the motor assembly process:

- Provide seated workstations to eliminate or reduce leg, back and arm fatigue.
- Provide an adjustable work surface (raise and tilt adjustability).
- Provide arm rests.
- Implement employees job rotating to jobs which do not require standing.
- Increase the frequency of job rotation and reduce the duration of time spent in each job.
- Modify the work-rest cycle to provide additional breaks.
- Install a platform for shorter workers.
- Provide ergonomically compatible floor mats.
- The most effective, but costliest, recommendation would be to automate this job.

REFERENCES

Burke, M. (1992). *Applied Ergonomics Handbook*. Ann Arbor: Lewis Publishers.

Eastman Kodak Company. (1983). *Ergonomic Design for People at Work*. New York: Van Nostrand Reinhold.

Ivergard, T. (1989). *Handbook of Control Room Design and Ergonomics*. New York: Taylor & Francis.

Sanders, M. & McCormick, E. (1987). *Human Factors in Engineering and Design*. New York: McGraw-Hill.

Sufalko, K. (1995, Nov./Dec.). *Safety, Environment and Health Professional Offer Advice and Interpretations for Your Questions about Compliance Standards*. Compliance Magazine. (p. 4).

Woodson, W. & Conover, D. (1964). *Human Engineering Guide for Equipment Designers*. Los Angeles: University of California Press.

CASE STUDY QUESTIONS

1. What were the ergonomic problems of concern in this case study?

2. What activities were performed to pinpoint the cause of the ergonomic problems?

3. What body parts were affected by stressors found in this work environment?

4. Identify solutions that you would recommend to eliminate or control the ergonomic problems identified in this case study.

ERGONOMIC PROBLEMS IN THE PRODUCT DESIGN INDUSTRY

Brad Eudy

INTRODUCTION

The organization of interest in this case study is a product design and development company that specializes in rapid prototyping and low volume, short run manufacturing. One aspect of the operation is rapid casting (prototype injection molding) using manual means. This was the area of concern considering the labor intensive processes involved in this operation. The casting area was set up as a laboratory, very similar to an institutional kitchen. The process involved lifting and placing a mold on a work surface; cleaning the mold with solvents; pouring, weighing, and mixing the liquid plastic; injecting the mold; lifting and placing the mold; lifting and placing the mold in a tank; tightening the tank closure bolts; removing the mold; opening the mold; removing the part and recycling the mold.

STATEMENT OF THE PROBLEM

The problem involved designing a new work area that complimented the process, fit the personnel working in the area, minimized physical effort, and provided an atmosphere and environment conducive to mental and physical health and safety.

Factors that were considered included:

- Maximizing usable space
- Appropriate work surface height
- Adequate illumination
- Proper ventilation
- Minimizing required lifting and transporting of materials
- Addressing standing work
- Reducing repetitive motions

CASE STUDY

The work surface was normal countertop height. This, with the average mold height of four to six inches, increased the work surface height to an awkward level. In addition, the casting technician is below average height (5'6"), which contributed to the work surface height problem. The old casting area was an interior room with no natural light and relied exclusively on UV fluorescent lights. This was a problem because of color matching requirements for the plastic parts.

Ventilation was a major concern for the new area. The previous area had a single local exhaust ventilation hood that inadequately removed solvent vapors, aerosol mold release agents and urethane vapors. The work surface was a long distance from the tanks that required not only lifting, but carrying of the molds to the tank. The floor area had a poured concrete floor with no shock absorbing mats. In addition, closing the tanks required turning up to 12 wing bolts, by hand, until the tank lid sealed.

RECOMMENDATIONS

Maximizing Space — Storage areas were created above the work area for the liquid urethane containers. Five gallon buckets were replaced with five-gallon containers with a spigot valve for dispensing. This eliminated the lifting of the containers every time a mold was filled (15 container lifts per day to one lift for every five gallons of material used for refilling activities).

The countertop and island table height was reduced from 38.5 inches to 32.5 inches to compensate for the casting technician's stature as well as the average height of the molds. This lowered the working height of the operation to an optimum working height for the operators hands (approximately 42 inches).

A large tray ceiling, tapered from a four foot square sky light provided additional natural light. In addition, direct lighting from halogen lamps illuminated the peripheral areas of the room. All fluorescent lamps were removed from the area. This lighting method aided in quality control activities including color matching and mold inspections.

A large local exhaust hood with side shields was installed to remove the majority of solvent vapors and mold release overspray. Exhaust duct lines were installed to vent urethane vapors from the back splash count area as well as to exhaust urethane vapors from the pressure tanks directly to the exterior of the building. By providing the entire work surface with local exhaust there was a decrease in the acquired movement of mold to areas that were ventilated. This significantly reduced lifting and carrying activities that would contribute to back and arm strain.

Workstations were placed closer together so that transportation required only lifting, turning and setting down motions. Lifting and carrying of molds or liquid containers was reduced by these modifications. Adjustable height carts were introduced to the work area to transport materials weighing over 40 pounds. A roller conveyor system was being designed for moving heavy materials on and off the carts.

In the original workstation layout there was no way to perform this operation without standing. Measures have since been taken to minimize stress on the legs and back. A cushioned rubber mat was placed in front of the workstations, toe kicks were constructed under the counter to allow the operator to get closer to the work without bending at the waist, and counter top width was reduced to prevent over extension for objects at the back of the workstation. Personal protective equipment allowances for appropriate footwear has been provided by the company. Running or tennis shoes with good support and cushioning have been recommended.

CONCLUSION

No data was available to measure the degree of improvement. Based upon employee comments, however, morale of the technicians has improved. No occupational injuries or illnesses have occurred since implementing the workstation changes. However, cycle time and quality have improved as a result of illumination and materials flow. The best evaluation would be a continuation of a perfect safety record.

CASE STUDY QUESTIONS

1. Identify the ergonomic hazards associated with the injection molding operation described in this case study.

2. List the ergonomic factors that were considered and provide examples for each factor mentioned.

3. What were the recommendations and solutions identified by the author of this case study? What recommendations would you make to control the ergonomic problems identified in this case study?

CHAPTER 2.6

MERCANTILE CASE STUDIES

The ergonomic case study found in this section is unique in that it examines an ergonomic problem in a "public" location. We often do not think about ergonomic issues associated with the construction and design of public facilities. Hand rails, steps and ramps, force requirements associated with opening and closing doors, and the width of corridors and entrance ways are just some of the numerous and very important ergonomic issues associated with facility design. With the passage of the American's with Disabilities Act, ergonomic analysis has become an increasingly important area for both industry and the public sector. Consider researching this area of ergonomics to increase your understanding of the health and safety issues associated with building layout and design.

After reviewing the following case study, identify the problems of concern, list the strategies employed to address those problems, and develop a list of actions that you would take to improve upon the recommendations presented. In addition, consider performing an ergonomic analysis of a facility of interest to you. Consider the physical characteristics of the facility and determine if modifications would be required to accommodate special populations including the handicapped and elderly.

ERGONOMIC FORCE REQUIREMENTS
TO OPEN MINI-MALL DOORS

Margaret Volney

INTRODUCTION

This ergonomic case study is based on a 4,624 square foot mini- mall located in the southeastern United States. The establishment consisted of 10 different retail and service oriented shops and employed 10 salespeople. The customer base of this mini mall was between 50 to 150 customers a day, 6 days a week.

The diversification of these shops contributed to 40 percent of the customer base being mothers with strollers, elderly people, and physically challenged individuals. Because one of the shops supplied delivery services, employees were routinely carrying out parcels, as well as receiving large shipments on a daily basis. The main entryway of this place of business, where all of this activity takes place was the focus of this case study (refer to Figure 2.6-1).

ERGONOMICS PROBLEM STATEMENT

This case study evaluated the ergonomic issues associated with two glass entry doors, specifically, the force requirements necessary to push or pull the entry doors open, relative to the business' clientele and service personnel that used them. A considerable number of complaints from customers had been recorded regarding the excessive force requiree to open the entry doors. Similar complaints have been made by delivery personnel who found it difficult to enter and exit when carrying large parcels and maneuvering hand carts. This problem affected everyone who accessed these doors. Thus, it was a focus of concern to correct this ergonomic problem before an accident occurred or before customers were lost due to the incon-venience of the entrance way.

CASE STUDY/PROBLEM ANALYSIS

Mechanical factors

The problem was the entrance way, which consisted of two glass doors encased in metal frames, with top door jambs installed with closer mountings on the push side of the door (refer to Figure 2.6-2 and 2.6-3).

Figure 2.6-1: A view of the two glass entry doors from out side the mini-mall.

Figure 2.6-2: A view of the two glass entry doors from inside the mini-mall. Note the closer jamb located in the top left corner of the photograph.

Figure 2.6-3: A close up view of the closer jamb located on the top left corner
of the glass door, mounted on the interior side of the entrance.

The previous use of this building warranted the back-check (BC)
adjustment valve (on the closer jamb) to be set at a high closing power
position. The force requirement to push or pull these doors has increased
over a 27-year period due to dirt and grime accumulation in the fittings, as
well as the BC valve becoming frozen in the maximum closing power
position.

Human factors

According to Woodson (1981) in the "Human Factors Design
Handbook," doors that are heavy were particularly difficult for women, older
persons, and handicapped individuals to use. In addition, he stated that adult
females were only about 2/3 as strong as adult males. In terms of age,
Hettinger (Grandjean, 1988) asserted that individuals have their maximum
strength between the ages of 25 and 35, and they lose approximately 25
percent of their strength by the time they reach age of 65. In relationship to
this mini-mall, the ratio of women to men customers was 3 to 1, and with a
large portion of its customer base being 65 or older. This emphasized the
need to correct this problem.

EVALUATION

It was determined that the approximate horizontal force requirement (for example, exertion forces toward and away from the body) to push open the glass doors at the mini mall was 28 pounds of pressure, and the approximate pull requirement was 32 pounds of pressure. Given the information on push/pull factors, the statistics concerning physical strength (refer to the previous section, "Case Study/Problem Assessment-Human Factors"), and the fact that the limb position and the direction of force application were important variables in determining the amount of force an individual could apply (Woodson, 1981), a chart was devised (refer to Table 2.6-1).

Much of the existing data on standardized strength measurements were based upon college and military men; these statistics do not reflect older or female individuals (Eastman Kodak Company, 1983). Thus, Table 2.6-1 was calculated based on healthy young males. This chart combines studies by Woodson and Conover (1964) and Grandjean (1988) on how the pushing force was greater than the pulling force at most positions of the arm while an individual was standing.

Position	Push (in pounds)	Pull (in pounds)
90° limb angle/male	150	120
90° limb angle male (35 to 65 yrs.)	113	90
90° limb angle male (over 65 yrs.)	85	68
90° limb angle female	100	80
90° limb angle female (35 to 65 yrs.)	75	60
90° limb angle female (over 65 yrs.)	50	45
60° limb angle male	127	125
60° limb angle male (35 to 65 yrs.)	96	94
60° limb angle male (over 65 yrs.)	74	71
60° limb angle female	86	84
60° limb angle female (35 to 65 yrs.)	64	62
60° limb angle female (over 65 yrs.)	48	46
Doors at site in closed position	28	32
Standard doors in closed position	15	16

Table 2.6-1: Arm strength and push/pull capabilities at selected degrees of flexion and selected age ranges.

It also pointed out that the pulling force was slightly higher at the 60° angle where the elbow was flexed, but that both pushing and pulling forces were greater in the vertical plane position. This table also exemplifies Grandjean's (1988) studies of the decreased amount of strength in both females and elderly individuals. Other factors that would influence and alter these statistics of push and pull horizontal forces, especially applied between the waist and shoulder levels, were (Eastman Kodak Company, 1983):

- Body weight
- Height of force
- Distance of force application from the body, or the amount of trunk flexion or extension
- Frictional coefficient of the floor
- Frictional coefficient of the shoes
- Duration of force application or the distance moved
- Availability of a structure against which the feet or back can push or prevent slippage
- Health
- Age in excess of 65 years of age
- Carrying cumbersome objects
- Pace
- Posture

Regarding the pull force requirements of these doors, a strong grasping grip was also necessary. This fact greatly influenced the pulling capability of people with arthritis, arm injuries, or those carrying packages or small children. Woodson (1981) further stated that an individual's grip declines about 16.5 percent by the age of 65.

RECOMMENDATIONS/SOLUTIONS

The following solutions were investigated concerning this ergonomic problem:

1. New closer jamb installed on existing doors (refer to Figure 2.6-3).
    a. Cost approximately $100.
    b. Change push/pull factor to approximately 15 pounds of required pressure.
    c. Drawback—unaccommodated customers and personnel.
2. Push pad hand-operated handicapped accessible doors
    a. Cost approximately $2,300.
    b. In standard operation, the doors would provide all of the benefits an ANSI/BHMA grade 1 door closer.

c. In power-assist operation, the doors could be opened with 1.5 to 5.0 pounds pushing force.

d. Drawback—ADA Manual requires 60 inches minimum turn around space in front of door in closed position. Existing site has 48 inches of turn around space.

3. Fully automated doors and standing air pad electronic doors

a. Cost between $4,000 and $4,500.

b. Automatically open when activated.

c. Drawbacks—Small children could activate doors. Longevity approximately 10 years. Loss of power causes doors to be in locked position (New standards are being implemented).

Action taken by management

All of the options and cost assessments that were made available to the company concerning this ergonomic problem were analyzed and reviewed. Because it was a small establishment, most of the solutions were cost prohibitive. Although management did not decide upon the most optimal choice, they chose a solution that would meet their needs and still remain within their budget. The decision was made to have a new top jamb closer mounting installed on the glass doors, since the frozen BC valve on the original closer was the reason the push/pull force requirement was so high.

REFERENCES

Eastman Kodak Company. (1983). *Ergonomic Design for People at Work,* Volume Two (pp. 386-389). New York: Van Nostrand Reinhold.

Grandjean, E. (1988). *Fitting the Task to the Man,* Fourth Edition (pp. 22-25). London: Taylor & Francis.

Woodson, W. E. (1981). *Human Factors Design Handbook* (pp. 70, 258-274, 772-776). New York: McGraw-Hill Book Company.

Woodson, W. E. & Conover, D. W. (1964*). Human Engineering Guide for Equipment Designers,* Second Edition (pp. 2-112-114, 5-35-36). Berkeley: University of California Press.

CASE STUDY QUESTIONS

1. Who were the individuals of concern in this case study? Why are the people involved in this location a factor that had to be considered in this ergonomic problem?

2. What contributed to the difficulty of opening and closing the door? How is this problem similar to problems observed in the industrial setting?

3. How do anthropometric and biomechanical factors affect the ergonomic problems in this case study?

4. What measurements were taken in this case study? Did these measurements indicate an ergonomic problem? Why?

5. List and briefly describe the factors that influence push and pull capabilities and limitations.

6. Identify the solutions recommended in this case study. What is interesting about these recommendations?

7. What recommendations would you make to address the ergonomic problems presented in this case study? What would you have done to increase the likelihood of support for your recommendations?

CHAPTER 2.7

OFFICE ENVIRONMENTS CASE STUDIES

The case studies found in this section examine ergonomic issues in office environments. While Video Display Terminals (VDTs) appear to be the office equipment that contributes to ergonomic risks in this environment, there are many types of office equipment that require repetitive motions. Calculators, communication switchboards, and cash registers are just some of the many types of equipment that should be analyzed from an ergonomic perspective. In addition to office equipment, one of the case studies in this section examines ergonomic issues associated with office layout. People, furniture, equipment, and general layout are all factors that require ergonomic analysis.

After reviewing the following case studies, identify the problems of concern in each one, list the strategies employed to address those problems, and develop a list of actions that you would take to improve upon the recommendations presented. In addition, consider performing an ergonomic analysis of an office facility or computer laboratory of interest to you. Consider the physical characteristics of the facility and determine if modifications would be required to accommodate "normal" employees as well as special populations, including the handicapped and elderly.

ERGONOMIC COMPLAINTS REGARDING
VIDEO DISPLAY TERMINALS

Leslie C. Ridings

INTRODUCTION

A paper plant that produces feminine pads and adult undergarments was very involved in health promotion and wellness, as well as safety and Ergonomic issues. There were 594 employees working for this plant 162 women and 88 men worked in the office areas of this plant. This study focused on the 250 personnel that specifically worked with VDTs.

In 1993, the company had seven formal complaints about employee wrist pain and one formal complaint about pain in the right hand. Many other informal complaints from office personnel about pain in their wrists, hands, shoulders, back, legs, neck and eyes were being brought to the attention of the nurses during medical screenings. The safety team and occupational health nurses immediately responded to these complaints because of the drastic long-term effects these potential problems could have on employee health and health care dollars. Long-term musculoskeletal medical problems associated with these complaints, such as carpal tunnel syndrome, tenosynovitis, and tendonitis, made the prompt response to this problem important.

Since many of the complaints were from employees that worked in an office setting, the occupational health nurses and the safety team decided to implement an Ergonomic program related specifically to VDTs. This program was implemented to teach employees how to layout and design their workplace using ergonomic concepts.

Ergonomics is defined as the study of human characteristics (capabilities and limitations) for the design or redesign of the living and working environment. In other words, ergonomics focuses upon adapting the workplace to the workers. A common health hazard afflicting office workers today is called cumulative trauma disorder (CTD). Cumulative trauma disorder is defined as "the excessive wear and tear on tendons, muscles, and sensitive nerve tissue caused by continuous use over an extended period of time" (Knoll Group, 1993).

The Bureau of National Affairs has reported that:

- In 1990, $27 billion was spent on CTDs. This includes medical care and lost income.
- U.S. corporations face more than 16 million lost workdays each year as a result of CTDs.
- More than half of all recordable injuries in 1989 and 1990 were ergonomically related.
- In 1989, the Bureau of Labor Statistics reported that 52% of all occupational illnesses resulted from CTDs.
- 40 million people who work with VDTs have suffered an unprecedented increase in CTDs.

In addition to increased awareness by industry, employees, and government in reporting ergonomic related disorders, the frequency of these injuries were due to changes in work processes and technology that expose employees to increased repetitive motion risk factors, resulting in increased costs (Rogers, 1994). This has brought about the need for finding effective solutions to manage ergonomic hazards and problems in the workplace.

CASE STUDY

The main concern of the company was to find an effective solution to decrease office worker complaints. In 1994, the health and safety team, along with the occupational health nurses, decided to implement an ergonomics program addressing VDTs. University ergonomists and physical therapists from a biomechanics laboratory and physical therapy clinic were brought in for professional help with these assessments.

The first step that the health and safety team initiated was the implementation of an Ergonomic program. The team decided to teach a class covering ergonomics and VDTs with professional assistance from the consultants. A questionnaire was developed for the office employees concerning their specific terminals. Some questions asked were:

Y N   My office area is so small that I have the feeling of being closed in (claustrophobia).
Y N   My files/folders are placed so that I always have to reach further than arm length.
Y N   The overhead lights cause reflection/glare on my computer terminal(s).
Y N   Temperature is often too low/high.

Y	N	My chair is without rollers, height adjustment, curved front edge, or lumbar curve.
Y	N	Keyboard is flat and not angled.
Y	N	I have a document holder but use it rarely because it is inconvenient.
Y	N	My wrists are bent up/down when I am typing.
Y	N	My back is hunched when I view the terminal.
Y	N	After work at home, many times I have my neck/back muscles tensed up.
Y	N	Overall, I have the sense that I am tensed up when I work on the computer.

This questionnaire was used to determine the problems at each terminal. The team then taught the employees how to change their workstations to eliminate these ergonomic problems.

During the second step a checklist was given to the employees to use when they returned to their work areas to assess their own terminals. This checklist helped the employees understand basic concepts of ideal office ergonomic features. Some topics emphasized in the checklist were:

- Ideal work space
- Overhead lighting
- Noise
- Proper work space layout and design
- Proper posture for the body

After the questionnaire and checklist were given, photographs were taken, using a 35 mm camera, of the terminals before and after Ergonomic intervention. The health and safety team then examined all of the office stations for proper settings.

RESULTS/RECOMMENDATIONS

The review by the health and safety team identified 12 ergonomically incorrect workstations thought to require help from professionals. Ergonomic consultants were asked to assess the situation. Of these 12 workstations, three were of particular interest to them. These three were thought to require prompt changes to reduce ergonomic risk factors. The following changes were implemented:

- Overall posture changes were recommended by adjusting chair/desk height.
- Chair height was reduced so that feet remained flat on the floor.
- Lumbar curve was established by changing the posture and/or chair.
- Front edge of the seat should not cut in back of knee area.
- Keyboard heights were raised to improve wrist and elbow position.
- Wrists on edge of table could cut off circulation to forearm/hands. Wrists were kept in a neutral position by use of a wrist support.
- Monitors were elevated so that the upper edge was at eye level.
- The work spaces were rearranged so that commonly used items were easily accessible and no unnecessary twisting, bending, or stooping was being accomplished.
- All document holders were lined up between the keyboard and the monitor and was the same height as the monitor in order to reduce back and neck pain, as well as eye strain.
- Glare screens were considered for monitors to reduce glare on monitors.
- Lighting needed to be at the side of the terminal instead of directly above the terminal in order to reduce glare on the screen.
- Headset for the phone was considered.

Stretching exercises were also implemented to train employees how to relieve stress in their hands, wrists, neck and back areas. Copies of these stretches were given to the employees for future reference.

CONCLUSION

Overall, everyone was pleased with the new layouts of their workstations. The company ordered ergonomic features such as wrist rests, glare screens, foot rests, and mouse pads for all the employees that needed them. When asked questions about their workstations and the changes that were made, the only complaint received was they would like faster computers and the company is working on that request!

The company's health and safety team realized that the VDTs were a serious problem for their employees. Continuous follow-up of workstations and Ergonomic training are recommended to help the employees continue to understand the basic concepts.

The application of ergonomic principles to the design of workstations, tools, and tasks may be the most significant contribution that companies can give to their workers to improve their overall health, satisfaction in their workplace, productivity, and overall safety. Cumulative trauma disorder is a major problem associated with VDTs. It is caused by three major factors: *repetition, awkward postures or positions, and excessive force*. These risk factors can be controlled by adjusting the workstation, varying work positions, reducing continuous or repetitious action, and by periodically stretching throughout the day.

This company has been concerned with the health of their employees and have realized that ergonomics plays a major role in each of the employee's health and safety. The recommendations for the new setup of the office personnel's workstations and the new ergonomic wrist rests and other Ergonomic tools installed have shown employees that the company does care about them and that efforts for correcting ergonomic problems could be achieved through a team effort.

REFERENCES

Dickerson, OB & Baker, WE. Practical Ergonomics and Work with VDTs. *Occupational Health Nursing:* 428-443, 1995.

The Knoll Group. *"Office Ergonomics: Working in Comfort."* 1993.

Rogers, Bonnie. *Occupational Health Nursing: Concepts and Practice:* 140-149, 1994.

CASE STUDY QUESTIONS

1. What were the physical complaints associated with the ergonomic problem in this case study? What are the typical health issues associated with VDTs? Were these physical complaints similar to those frequently mentioned in the literature?

2. List the CTD statistics cited in this case study.

3. How did this company address the ergonomic issue of VDTs in this case study?

4. A survey was conducted to determine where the ergonomic problems were occurring. What questions in this survey do you consider important in pinpointing the extent of the ergonomic problem?

5. How would you use the results obtained from a survey like the one presented in this case study? What additional questions would you want to ask?

6. A workstation evaluation checklist should include what factors of importance?

7. What recommendations do you consider important in this case study?

8. What would you do to address a VDT issue in your organization?

AN ERGONOMIC PROGRAM FOR SOLVING
VDT WORKSTATION PROBLEMS

Steve Joyner

INTRODUCTION

This case study addresses the measures taken by a large service industry to reduce repetitive strain injuries (RSI) associated with worker's who utilize VDTs. These workers repeated tasks over and over which could lead to serious strain and overall health concerns.

Repetitive motion disorders are common occurrences among VDT workers, assembly line workers, and other office personnel whose jobs entail repetitive movements. Carpal tunnel syndrome and tendonitis, are just two of the many possible complication associated with repetitive motion tasks.

This company employed over 7,000 research and development personnel who utilized VDT workstations. Some of the tasks associated with VDT workstations, as well as repetitive strain injuries, are long, uninterrupted keyboard sessions, continual reaching, stretching and twisting, and also continual viewing of video monitors. After receiving numerous complaints and worker's compensation claims involving RSI and similar injuries, this company was prompted into taking the following approach in order to reduce these repetitive strain injuries associated with VDT workstations.

CASE STUDY

First, the health and safety staff concentrated on identifying the RSI risk factors associated with VDT tasks. This began by reviewing existing data from interviews with the injured and by evaluating workstations and employee's personal work habits. Workstation furniture and equipment was found to vary from modern ergonomic designs to outdated 1940's vintage tables and desks. The staff realized that wide keyboards and bulky monitors did not fit very efficiently and that accessories could not be mounted easily on some of the existing equipment. The conclusion from this initial analysis was that task repetitiveness and awkward postures were the primary risk factors.

Next, the team initiated a joint effort between the medical and safety staff to create a medical protocol. This protocol consisted of medical evaluations of employees with work-related and work aggravated musculoskeletal complaints. The company's medical staff carefully analyzed employees' symptoms to determine whether or not they were related specifically to repetitive movement or to existing medical problems. Employees were carefully monitored to determine what parts of their work might have caused the ergonomic problems. The medical staff provided advice to workers concerning their personal work habits and made suggestion for corrective actions.

With the data form in hand, the medical and safety staff set out to perform a worksite task analysis. The purpose of the analysis was to correlate medical problems with work-related causes and to specify corrective actions. The staff reviewed workstation layouts, furniture and worker habits. They found that employee work habits differed in some instances, such as typing methods, which were difficult to control. However, many of these habits were significant factors that caused RSI. Most importantly, they concluded that all of the workstations needed to be modified. Placement of monitors, keyboards, chairs, and other accessories would all be taken under consideration for modifications.

Finally, the team began a modification process. The most successful modifications were the simplest and least costly. The VDT furniture was retrofitted with accessories to improve body posture and body symmetry. Keyboard supports, capable of side to side movement and swiveling, were incorporated to reduce hand and wrist strains. Adjustable mouseboards that brought the mouse closer to the user were added to reduce reaching. Palm rests for keyboards, adjustable chairs with modern ergonomic features, and "video display task glasses" were all part of the modifications. "Video display task glasses" were simply physician prescribed bifocal glasses that were designed for the viewing distance between the user and their video monitors. These glasses were prescribed as needed to reduce eye strain.

CONCLUSION

After all of the workstation modifications were complete, recommendations for individual training given, and eyeglasses prescribed, employees were given a 28-day trial period. They were then asked to complete a questionnaire. This data and feedback was a continual mechanism used for measuring overall effectiveness for the ergonomic modifications. Feedback, thus far, suggested that the modification averaged approximately $300 per station. The health improvements appeared to outweigh the costs associated with these modifications. Improved employee relations, improved productivity, and most importantly, a significant decrease in employee health claims were some of the improvements realized.

CASE STUDY QUESTIONS

1. List the tasks that were associated with repetitive motion problems identified in this case study.

2. Base upon initial findings in this case study, what were the hazards that contributed to the ergonomic problem?

3. How was the medical group involved in this project? If your organization did not have an in-house medical department, what might you do to obtain medical input?

4. Worker habits were identified in this case study as a possible contributor to ergonomic problems. Develop a list of personal habits that you perform, or that you have seen performed, that could contribute to ergonomic problems while using VDTs.

5. What were some of the solutions that were implemented to correct or eliminate VDT related ergonomic problems?

6. What would you do to eliminate the problems identified in this case study?

AN ERGONOMIC SYSTEMS APPROACH FOR SOLVING VDT WORKSTATION PROBLEMS IN THE AGRICULTURAL INDUSTRY

James Kohn

INTRODUCTION

A Certified Professional Ergonomist (CPE) was hired by an agricultural cooperative to assess ergonomic hazards at its corporate headquarters. An ergonomic audit instrument was used to collect computer workstation data.

The audit revealed numerous ergonomic hazards. Two contrasting conditions existed that contributed to these hazards. In one instance, some of the locations had computers placed on traditional office furniture and desks not designed to accommodate computers. In other workstation environments, new chairs and tables were provided. However, the "ergonomically designed" furniture had not been adjusted to fit the operators using the workstations.

The majority of offices audited had computers that were placed on traditional office furniture. In addition, older style chairs were in use at those locations resulting in an undesirable condition. These chairs failed to provide adequate back support and the height adjustment for effective worker-keyboard interface was nonexistent.

CASE STUDY

As a result of the audit that was conducted at this location numerous ergonomic hazards were identified. Common ergonomic hazards that were found in these locations included:

1. Stands, tables, supports for the keyboard that were not adjustable. With the use of standard office desks, body posture during computer operation was unsafe. This resulted in dorsiflexion (the movement of the hand that decreases the angle between the back of the hand and the arm) of the operator's wrists increasing the likelihood of repetitive motion injuries among operators required to perform data entry or word processing activities for prolonged periods of time.
2. Computer monitors were set at improper angles. In some instances, computer monitors were placed upon system units resulting in an operator head posture which facilitated neck, shoulder, and back fatigue.

3. Arm, head, and feet supports were not available. In a majority of the observed workstations, arm, hand, and feet supports were not used. This resulted in improper posture increasing the potential for physiological fatigue in the short term and possible injury in the long term.

4. Monitors were located in front of windows or immediately below and forward of overhead lighting fixtures. As a result of improper equipment placement, glare related conditions existed for equipment operators. This condition could be expected to result in eye fatigue and strain.

5. Work area surfaces were inadequate and document holders unavailable or poorly placed when available. In some instances, working surfaces or document holders that were used resulted in operators bending, twisting and turning their necks. This could be expected to increase neck, shoulder and back fatigue.

6. Chairs that failed to provide adequate ergonomic support for computer operators. In many instances, chair back and shoulder support was not available and chairs could not be easily adjusted to provide necessary support.

7. Operators could not properly adjust tables and chairs. In many instances, operators set up make-shift computer equipment layouts, including placing keyboards in open desk draws or on laps. These conditions would increase the probability of fatigue and physiological strains.

Ergonomic problems existed in office locations that had adjustable furniture as part of the computer workstation. Several offices audited had computer workstations with "state-of-the-art" office furniture. Despite the availability of good equipment, undesirable conditions existed from an ergonomic perspective. Common ergonomic hazards in these locations included:

1. Chairs and keyboard stands incorrectly adjusted or improperly used for the workstation operator.
2. Computer workstations placed in locations that promoted screen glare.
3. Operators did not correctly adjust blinds to reduce glare and screen contrast problems.
4. Operators unaware of proper computer workstation operating positions and how to correctly adjust chairs, keyboard supports, stands and tables.
5. Ideal placement of documents to reduce neck and upper torso movements.
6. Strategies that can be used to reduce repetitive motion related problems, including but not limited to exercise, change in activities and periodic breaks.

RECOMMENDATIONS

The consultant recommended the following strategy to reduce the risk of ergonomic related hazards:

1. Purchase foot and arm/wrist rests to promote proper posture among operators requiring additional physiological support.
2. Prioritize computer workstation location by percent time spent on the units by operators. Replace chairs and furniture with sound ergonomically designed equipment in those locations where the most significant risks exist.
3. Consider job enrichment activities for computer operators who spend more than 50 percent of their time on keyboard related tasks. Employ alternative tasks to minimize time spent in repetitive motion activities.
4. Ensure adequate breaks to allow computer operator stretching and exercise activities for the reduction of fatigue and stress.
5. Evaluate all computer workstations and modify workplace environments to minimize illumination related problems.
6. Evaluate all computer workstations to ensure that layouts comply with acceptable ergonomic guidelines.
7. Ensure the availability and use of document holders to minimize computer operator twisting and bending motions.
8. Perform periodic audits, looking for proper computer operator posture and recommend modifications necessary to minimize back, neck and arm strain.
9. Conduct ergonomic training for all computer operators.

CONCLUSION

This organization accepted the recommendations presented by the consultant including the ergonomic training classes. Forty-five minute ergonomic training classes were conducted for all computer users. The goal of the training program was to ensure participant knowledge of ergonomics, proper computer workstation layout and design, and repetitive motion injury prevention methods. The training program consisted of the following objectives:

1. definition of ergonomic terms
2. application to workplace and home equipment and furniture design
3. symptoms associated with poorly designed environments
4. discussion of specific concerns expressed by course participants
5. identification of ergonomic hazard countermeasures, and

6.   commitment by the course instructor to observe all interested participants and evaluate their workstations.

After all of the operators were trained, approximately 20 department representatives were involved in a train-the-trainer program. This was conducted to ensure that new employees would be trained in the principles presented by the consultant.

When each training session was completed, the consultant accompanied by a department representative, visited each workstation where training program participants worked. The intent was to help each operator adjust the furniture and reposition the workstation to eliminate ergonomic hazards. In addition, if changes could not be made, a list of required furniture or equipment (like wrist and foot rests) was compiled. This list was then submitted to management.

Of interest to the consultant was that although participants knew the correct way to set up computer workstations following the training program, they had no idea how to personally adjust their own furniture and equipment. The consultant and department representative then worked with each employee to correctly adjust their workstation.

Implementation of most of the recommendations and the ergonomic training program reduced the number of employee complaints by over 50 percent. Employee morale was reported by department supervisors to have improved and no cases of repetitive motion injuries were reported for an 18 month period following implementation of this program.

CASE STUDY QUESTIONS

1. In this VDT case study, what contrasting conditions existed that contributed to the ergonomic hazards?

2. List some of the ergonomic hazards observed in this case study for the two areas of interest.

3. Review the list of recommendations. Which recommendations are the most important ones for eliminating ergonomically related VDT hazards over the long term?

4. Identify the components of the recommended training program. What were the strengths and weaknesses of this training program?

5. A train-the-trainer program was part of the training approach employed in this case study. Is this a useful activity? Why?

6. Why was direct observation follow-up activities after training important for the success of this case study?

OFFICE ERGONOMICS AND ENGINEERING DESIGN

Mark D. Hansen

INTRODUCTION

Believe it or not, none of the office furniture on the market today is truly ergonomic. How can that be, with all of the products stamped "ergonomic" on its side? Well, in many cases, that is all they are as far as ergonomics is concerned. It is the same old furniture with a stamp saying "ergonomic" on the side. It is merely a ubiquitous marketing tool. I call them the "Ergonomics R Us" vendors. However, there are companies that sell office furniture that is *adjustable*. But ergonomics goes much farther than that. Even though office furniture is in some cases adjustable, unless it is adjusted to the person using the workstation, it is not ergonomic.

During the 1970's, modular furniture swept the country based on the claim of adjustability. Taking a closer look at that claim reveals that when a personal computer and a monitor are placed on the modular desktop the **ease** of adjustability is severely diminished. Couple this with a mobile work force, one that often relocates workers every six months to a year within a facility, and adjustability virtually disappears. Why is that? First of all, most workers don't know (or care) that their office furniture is adjustable. Second, they think that this is my office and I must live with the layout. The adjustability is not obvious to the uninitiated worker. Even when a company takes great care in configuring the workstation correctly, when workers re-locate the problem begins all over again.

Many of the solutions implemented to solve ergonomic problems fall prey to this often invisible problem. This isn't even the worst case. What if you are trying to implement ergonomics for a customer and they are using 20-year old office furniture and don't plan on upgrading their furniture in this decade? How do you convince them that the investment will save them time, money, productivity, and payoff almost immediately?

Doing ergonomics pays off

Ergonomics pays for itself in many ways. Quantifiable productivity increases have been seen anywhere from 5 to 50 percent. Decreased absenteeism, increased worker satisfaction, reduced error rates, and decreased worker's compensation are just a few of the benefits and cost savings of implementing ergonomics. In this day and age of "competitive advantage," ergonomics is playing a greater role. It just makes good business sense to implement ergonomic workstations that are easily adjustable.

Avoiding exposure to the uncontrollable costs

Office workers are candidates for work-related diseases, the most widely known is cumulative trauma disorders (CTDs). CTD, simply put, is overuse and overexertion syndrome. Overuse refers to repetitive motion, while overexertion refers to excessive force. Most office workers are exposed to overuse rather than overexertion. Because the cost of CTDs today are so exorbitant, the costs of not implementing ergonomic document imaging systems can virtually cripple a business. The cost of CTD on a per worker basis can be as high as $50,000 to $75,000. This is an end-to-end cost that includes diagnosis, therapy, surgery and rehabilitation. Businesses have spent billions of dollars on CTDs alone — CTDs from people working at video display terminals, some of which were probably document imaging systems.

Epidemiology

It must first be conceded that people rarely die as a direct result of injuries and diseases involving ergonomics. Thus, mortality statistics are not relevant to the question of reducing ergonomic diseases and injuries in the workplace. It is important, however, to realize that the health and quality of life is greatly reduced for a large proportion of the population because of acute and chronic ergonomic diseases and injuries. The following items are from the Bureau of National Affairs, Occupational Safety & Health Reporter.

- In 1990, $27 billion was spent on CTDs. This includes medical care and lost income.

- U.S corporation face more than 16 million lost workdays each year as a result of CTDs.

- In 1988, more than $10 billion (about a third of worker's compensation) was paid for repetitive motion injuries.

- The number of reported cases of repetitive motion has increased eightfold since 1981 (See Table 2.7-1).

- More than one-half of the recordable injuries in 1989 and 1990 were ergonomic-related.

- In 1989, the Bureau of Labor Statistics reported that 52% of all occupational illnesses result from Cumulative Trauma Disorders (CTDs). The CTD incidence rate has increased more than 3,000% among certain clothing industry sectors and more than 2,500% among automobile workers.

- Forty million people who work with visual display terminals have suffered an unprecedented increase in CTDs.

- North Carolina reported an increase in CTD cases reported by more than 6,000% in 1989 from 1986. North Carolina's overall incidence rate in 1989 was 8.2 injuries and illnesses per 100 full-time workers. The national rate is 8.6.

From this data alone it should be clear that ergonomic related injuries are a rapidly growing concern in today's office workplace. The cost and frequency of ergonomic related injuries easily justify the benefits of incorporating ergonomics into the workplace. But there is more.

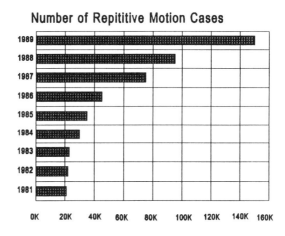

Table 2.7-1. Repetitive Motion Injuries, Number of Cases. (Bureau of Labor Statistics).

Social/legal.

I am convinced that we don't live in the Information Age; rather, we live in the Litigation Age. Whenever there is a problem, the solution is to sue one's opposition to get satisfaction. There is substantial social and legal support to incorporate ergonomics in today's workplace. The central social trend involves accommodating workers with diverse physical capabilities and disabilities. The result has meant that simple rules of denying people of a certain age, gender, race, or apparent disability of fair opportunity to perform a job are no longer valid (Miner and Miner, 1978). For instance, an allowable weight-lifting standard recommended in the middle 1960's by the International Standards Organization is no longer valid in many countries because it discriminates against women and people of advanced age by stating limits for these groups that are far below those for a younger male. It has been realized in recent years that a person's physical attributes, such as strength, flexibility, and endurance vary greatly within gender and age groups.

For these reasons it has become necessary to know with as much precision as possible the functional capacity of an individual relative to a job's performance requirements to assure that the person is not unjustly denied a job. This is also good medical practice and, indeed, is required by some occupational health and safety statutes. In this latter sense, the use of functional capacity tests can assist in assuring that a job applicant will not be over-stressed when placed on a new job.

For instance, recent studies have shown that people that were not able to demonstrate an isometric strength sufficient to adequately perform the strength requiring tasks in various jobs assigned to them had a threefold increase in both the incidence and severity of musculoskeletal injuries compared with their stronger peers (Chaffin, 1974; Chaffin, Herrin and Keyserling, 1978; and Keyserling, et al., 1980). The development and validation of appropriate functional capacity tests of both job applicants and people who are to seeking to return to work after suffering a musculoskeletal injury is a major occupational biomechanics effort.

Regarding legal support there have been numerous OSHA fines and individual law suits filed against employers. Some examples are as follows:

OSHA fines

- Samsonite — fined $1.6 million for the lack of ergonomics in the workplace.

- Cargill — fined $400,000 for the lack of ergonomics in the workplace.

Lawsuits

- Dennis vs. Communication Machinery Corporation. Five workers seeking $40 million each in compensatory damages and $20 million in punitive damages for ergonomic-related injuries.

- Aikman vs. Electronic Pre-Press Systems, Inc. Each of the plaintiffs are seeking $10 million for negligence and another $10 million for strict liability for ergonomic-related injuries. Their spouses are seeking $2 million a piece for the loss of consortium.

- Taylor vs. System Integrators, Inc. The plaintiffs are seeking $10 million in punitive damages and attorney's fees and costs.

Class action suits

Taking lawsuits to the next level, some people are getting together and filing class action suits against their employers (VDT NEWS, 1992):

Company	#	Company	#
IBM Corp.	67	NCR Corp.	21
AT&T	45	Northern Telecom	13
Computer Consoles	43	Unisys Corp.	18
Atex Inc.	36	Systems Integrators Inc.	12
Memorex Corp.	11	Wang Laboratories, Inc.	9

The above reasons should be sufficient to convince any company that implementing an ergonomics program is both good business sense as well as good for the work force.

Who should I get to make sure I do it right?

Within the past few years the Board of Certification on Professional Ergonomics, based in Washington, was formed. One of the certifications that they offer is a Certified Professional Ergonomist (CPE). Its Board of Directors consists of professionals with Ph.D.s in ergonomics. The practitioner population is ostensibly from the Human Factors and Ergonomics Society based out of Santa Monica, CA and was formed in 1956.

The minimum criteria to apply for this certification is an M.S. in a related field and seven years of applicable experience. Getting someone with a CPE doesn't guarantee that the job will get done right, however, it significantly increases the probability that it will. This philosophy is similar to that of other professions (for example, medicine (M.D.) and engineering (P.E.)). Someone who fits into this category should be able to meet the minimum requirements for implementing ergonomic programs.

The reason for this requirement is to increase the probability of success, and protect the customer and the integrator regarding the implementation of ergonomic programs. If you rely solely on hand therapists, physical therapists, occupational health nurses, someone off the street with a certificate from a three hour course, or anyone that has not been specifically trained in ergonomics, you are operating at risk.

How is ergonomics done?

The practice of ergonomics varies from incorporating principles in the design process of new systems to evaluating already existing facilities and making recommendations. In the former, the ergonomist determines the target user population, identifies the tasks the users will perform, and designs the environment for the user with consideration to the tasks. In the latter, the ergonomist may use one of many methods to assess the effectiveness of the workplace. Methods include interviewing various employees, inspecting the facility, examining injury and illness records, or using questionnaires to assess the comfort level of the workplace.

For example, questionnaires could help identify muscle groups that become painful during the course of a work day. From this data the ergonomist identifies the problem area, implements the change(s) (See Table 2.7-2), and re-evaluates the workplace as required. If the worker's task involves lifting, the ergonomist will generally use the NIOSH formula to determine how often the worker can perform that task during the shift. Additionally, this can lead to the identification of lifting aids to mitigate the effect on the worker (Snook, 1978).

Ergonomics issues

One of the issues impacting ergonomics focuses on potential injuries and the management perception of cause. The nature of ergonomic injuries is virtually hidden from the layman's eye. There is no overt "accident" that was ever witnessed. For example, the employee didn't fall down while performing a work-related task or get hurt using a punch press. Rather, the employee was injured quietly, while sitting at the VDT or at a manufacturing workstation performing a repetitive task.

As a result, there is no tangible manifestation of the injury. This leads to the logical reason management would question the injury. If the injury is not tangible, there is a tendency to believe the employee is "goldbricking," rather than being truthful, and that the employee is looking for a way out of the particular job. Changing this perception is paramount for progress. Education through all levels of staff and line organizations is essential to identifying this issue of ergonomics. Education may include the use of printed material, reminding employees to be ergonomic-minded, formal classroom training for employees and line and staff managers, and periodic re-training to facilitate retention, etc.

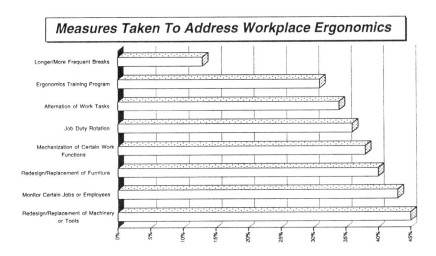

Table 2.7-2: Measures taken to Address Workplace Ergonomics. (Bureau of National Affairs, Occupational Safety and Health Reporter)

How about an example?

When examining computer workstations, the following components should be evaluated for positive ergonomic characteristics:

Chair: A chair should have a five point star base for stability, an adjustable backrest (angle, height, and depth) that provides lumbar support and an adjustable seat pan (height, forward and backward, and tilt angle). If arm rests are provided, they should be adjustable up and down, in and out and swivel (e.g., like a wrist rest). The edge of the seat pan should be at least 4 inches from the soft tissue area behind the knee. Controls for these features should be easily identifiable with their stated function, to reduce learning and enhance usability.

Document Holder: When the primary task is data entry, a document holder should be provided that is the same height and distance from the user as the display screen.

Display Screen: The top of the display screen should be slightly below eye level (20°).

Keyboard: The keyboard should be detachable and adjustable to allow straight/parallel hand-forearm posture. This is often accomplished using a wrist rest. The height of the wrist rest should equal the home row key height. Fingers on the home row of a keyboard should be approximately 0 to +1.5 inches above the elbow rest height. The keyboard slope should not be greater than 15°.

Desk or Table Top: The worksurface should allow leg room for posture adjustments for the seated worker while also providing a 90° angle of the elbow and the worksurface. The same is also true regarding the elbow angle for the sit/stand and the standing user.

Lighting and Glare: Workstation lighting should provide a 10:3 ratio. That is, characters should be 10 times brighter than the screen background. High luminance sources in the peripheral field of the VDT should be avoided. The VDT screen should be positioned to avoid glare. Glare and screen reflections can be eliminated by moving or tilting the terminal, and if necessary, by using an antiglare or polarizing filter.

Breaks: Vision and performance can be improved by allowing users to take a 15-minute break every 2 hours.

Posture: The head should be tilted 15° forward or less to maintain a vertical position. The elbows should be kept close to the body or supported. The lumbar curve of the back should be maintained. Feet should never be allowed to dangle and should always be supported.

Radiation Effects: Reproductive disorders, skin rashes, cataracts, and epileptic reactions have all been cited as possible radiation-related effects of VDT use. Even though there is no conclusive proof, these radiation-related matters all deserve more extensive scientific research. However, care should be exercised to place VDTs at least 28 inches from the user to avoid exposure. This may cause some screen re-design as many imaging applications are already at or below typical character height causing eye strain.

VDT Quality: There are three factors that affect VDT quality when used for office worker applications: resolution, screen refresh rate, and screen size. Resolution should be 1600 pixels x 1200 pixels (on a 19-inch diagonal) or about 120 dots per inch (DPI) to yield optimum readability. Screen refresh should be at or above 70 Hertz (Hz) non-interlaced. Screen size for one page formats (portrait) should measure at least 15 inches diagonally, while two page formats (landscape) should measure at least 19 inches diagonally.

Dry Eye: Recent studies have indicated that when people view monitors they blink less causing the eye to dry out. As a result, fatigue can set in and the need for more diversified non-VDT tasks implemented. This may call for more creative task design to ensure that the tasks assigned are still accomplished while not exposing users to undue risk.

Utilizing an ergonomist with the proper credentials and background, workstations can be set up to facilitate productivity of imaging system implementations with a high degree of confidence. The background to look for includes an M.S. in ergonomics or a related field with several years of experience. The credentials to look for include a Certified Professional Ergonomist licensure and/or a registered Professional Engineer (P.E.).

Ergonomics is a growing concern for industry and is expected to continue to grow in the 1990's. Attention to proper workstation design can not only eliminate or reduce ergonomic hazard exposures to workers but potentially redouble productivity of implemented imaging systems. To be competitive in the 1990's, maximizing productivity wherever possible will be a must.

CASE STUDY

This project was not like most projects. This project was extremely sensitive to the users due to all of the changes that were occurring in the Vermont Department of Motor Vehicles (DMV). The Vermont DMV was not just implementing a new system to help them do their jobs better. They were implementing a system that changed the way the worker interfaced with the customers, fellow employees, and the new automated system. Implementing this new system required re-designing and re-engineering the process, as well as completely changing the office furniture. Unfortunately, the office furniture had already been selected, which was modular furniture. The paper-driven process was being replaced by an electronically-driven process which included many changes to reduce the number of steps that were formerly done manually. The new system they implemented changed the whole culture of their job environment.

When changes like this occur, employees are generally resistant to change. People resist because they are quite familiar with the "old" way of doing business. Changing to a "new" way of doing business requires learning something new and different from what they are accustomed to doing. Instead of doing things manually, getting up and hand carrying forms to the next person to sign, they now do all of this in an automated fashion. They send these same forms electronically and they also sign these forms electronically, without a pen.

This drastic culture change caused anxiety in employees for several reasons. One reason for this anxiety was simply the change from one thing to another, regardless of the type of change. One reason was the fact that employees must learn something new that was slightly different from the previous way of doing business. There was a learning curve associated with learning this new process. With every learning curve there are mistakes, and every employee deals differently with mistakes.

This culture change also caused anxiety due to the fact that when tasks are performed manually, there was an assurance by the employees that this task was completed. This type of closure was not always inherently evident in the new automated system. The result was that employees had to learn to trust the system that it would perform as expected. Once this was achieved the employee anxiety waned.

Alliances

This project was different from most projects as the Vermont DMV assembled a transition user group. The user group was responsible for evaluating all of the changes that would affect the employees, identifying as many changes that enhanced the job environment, and most importantly, approving or not approving candidate designs that related to the new implementation.

This user group became the premier power group for the Vermont DMV. They had the complete support of top management. If this group did not like a particular item, it would not get implemented. On the other hand, if this group did like a particular item, it would get implemented. Forming an alliance with this group could easily spell success or defeat, depending on how they were approached.

As an ergonomics consultant attending their meetings, it was crucial to listen to their needs and design a job environment that met those needs. There were several ways of integrating ergonomic principles. One way was to latch on to meaningful concepts that become points of discussion during user group meetings. Once this was done the ergonomist can provide supporting information, statistical information, examples, war stories, etc. to illustrate the point. Another way was to supplant ergonomics principles into user group members so that it becomes their idea. To do this the ergonomist must release any pride of authorship. The goal was to implement an ergonomically-correct design, not boast of who won the battle.

Once an alliance was built with the user group, they become the champion for ergonomics. The ergonomist was merely one of the members of the team. As a team player, the user group began asking ergonomic questions about equipment. As a result, ergonomic equipment would be used in the design to the greatest extent possible.

Early involvement in design

Involvement in the user group began early in the conceptual phase for the Vermont DMV. Once acceptance was gained, implementing ergonomics principles into the design became second nature. One of the positive points of early involvement was that the ergonomically-correct design evolved from the early design concept. The ergonomically-correct design became the de facto standard. This involvement prevented a non-ergonomically-correct design becoming the de facto standard. The main benefit was that of not having to change a poor design and overcoming the institutional inertia of "the workstation is already designed."

Performing a trade study

Ergonomics consultants are rarely in the business of selling equipment to clients. This is due to an inherent "conflict of interest" between recommending equipment that an ergonomic consultant sells. The ergonomic consultant is clearly biased toward their own product(s) for obvious reasons. However, an ergonomics consultant could provide a list of vendors that meet the design requirements and environmental restrictions for workstations, chairs, etc. The best approach is to provide the client with a specification, a list of attributes, and a list of ranges of adjustability for equipment that is to be used in the design[1]. For example, a checklist for an ergonomically-correct chair for a seated environment would look something like the one in Figure 2.74-1.

Initial architectural design

The initial architectural design reflected the architect's design concepts with little input from the transition user group. This was the first flaw. The architect failed to discuss with the user group the basic needs and requirements of the workstation. The result was that the architect designed a typical seated environment with the DMV employees having to physically look up to the customers standing in front of the workstation while ordering new licenses (see Figures 2.74-2 and 2.74-3). A clear desire for the user group was to have employees look eye-to-eye or look down to customers. The reason was the subtle authority inference by being physically higher than the customers. An ergonomic buttress to this was the ability of employees to sit or stand at their workstation. Another problem was the placement of the computer. Employees must constantly look 90° from normal to view and interact with customers. These two design errors would severely predispose the employees to workplace dissatisfaction and perhaps even ergonomic injuries.

The second flaw, and probably more severe than the first flaw, was that the architect has already convinced top management to purchase modular furniture. The problem with modular furniture was that it is not user adjustable and most employees do not know it was adjusted. Once a workstation was loaded with several hundred pounds of equipment it is virtually unmanageable by a single employee, especially by females, which was the primary workforce for the DMV.

A third flaw was that all of the equipment was placed on a worksurface placed at 2 feet 9 inches above the floor. The problem was that employees had to reach over all of the equipment to interact with the customers that visit their booth. For example, the scanner was placed such that the paper had to be retrieved from the far side of the scanner forcing the employees to move around and reach a long distance throughout the day.

A fourth flaw related to the second flaw, which required a team of facilities people to go around and change workstation heights to fit employees when operators move from one workstation to another. The cost of buying user adjustable equipment would account for the cost difference between the two types of equipment.

RECOMMENDATIONS/SOLUTIONS

After working closely with the user group an ergonomically-correct solution was provided and approved by the user group and top management. The recommendations[2,3] (shown in Figures 2.7-4 and 2.7-5) were as follows:

1. Place the laser printer and the scanner below the worksurface so that the output paper is at worksurface height. This relieves the employees from reaching 12-16" above the worksurface for products.

2. Share the laser printer between two operators to relieve crowded worksurfaces.

3. Place the monitor on a mechanical arm above the worksurface and partially in the line of sight with the customers. This relieves the employees from repetitive head turning to talk to customers. Also, place a copyholder on the monitor to relieve head and eye movements during data entry. This solution also provides employees flexibility with the ability to place the monitor on the right- or left-hand side. depending on their preference. To do this, the monitors and the storage areas can be switched to accommodate this need.

4. Raise the workstation 10" to provide the employees to look eye-to-eye or look down to customers. The higher workstation also provides employees with the flexibility to sit or stand during the course of an 8-hour day.

```
┌─────────────────────────────────────────────────────────────┐
│                      CHAIR CHECKLIST                          │
│                                                               │
│  Number          Checklist Item                         Y/N   │
│                                                               │
│  1.   Does the chair have a five point base?             _    │
│  2.   Does the chair provide a backrest for lumbar support?  _│
│  3.   Is the backrest height adjustable between 6-9 inches? _ │
│  4.   Is the backrest width between 12-14 inches?        _    │
│  5.   Does the seat pan tilt forward and backward 20°?   _    │
│  6.   Is the seat pan width between 17 and 20 inches?    _    │
│  7.   Is the seat pan depth between 15 and 18 inches?    _    │
│  8.   Is the seat pan cushion at least one inch thick?   _    │
│  9.   Are the arm rests at least 2 inches wide?          _    │
│  10.  Are the arm rests at least 8 inches long?          _    │
│  11.  Are the arm rests adjustable between 7.5-11 inches above│
│         the sitting area?                                _    │
│  12.  Do the arm rests rotate?                           _    │
│  13.  Does the chair provide the freedom/ability to change   │
│         posture at frequent intervals?                   _    │
│  14.  Does the chair shape support alternate postures?   _    │
│  15.  Are the chair adjustments easily performed?        _    │
│  16.  Is the seat upholstered to reduce sweating?        _    │
│  17.  Does the chair have casters and a swivel to provide easy│
│         positioning?                                     _    │
│  18.  Is the seat cushion too "hard" or too "soft"?      _    │
│  19.  Is the chair stable on the floor?                  _    │
└─────────────────────────────────────────────────────────────┘
```

Figure 2.7-1. Example Ergonomics Checklist for a Chair for a Seated Environment.

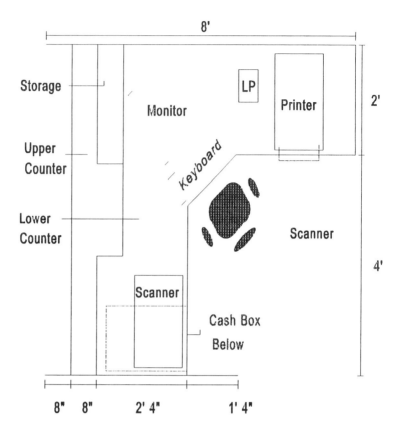

Figure 2.7-2: Top View of the Architect's Initial Design.

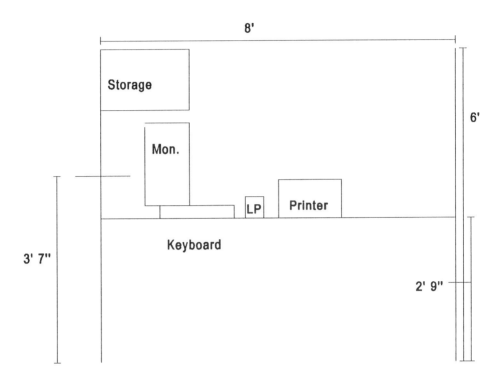

Figure 2.7-3: Side view of the architect's initial design.

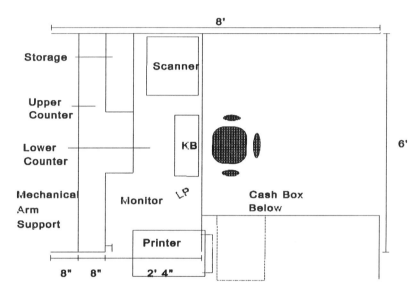

Figure 2.7-4 Top view of the ergonomic recommendations.

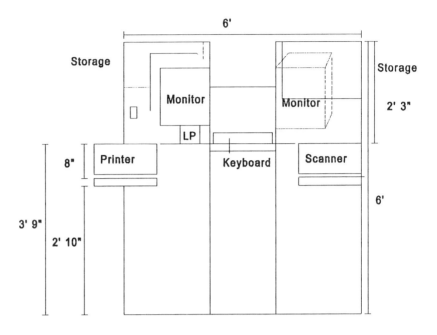

Figure 2.7-5: Side view of the ergonomic recommendations.

CONCLUSIONS

Successfully integrating ergonomic principles into the design of the work environment requires much more than merely knowing ergonomics principles. Successfully integrating ergonomic principles requires having the requisite interpersonal skills to be a team player, and knowing which group to ally with for support, promotion and implementation of ergonomics principles.

Successfully integrating ergonomic principles requires an understanding of the job environment where the workstation is being implemented. This includes performing a task and job analysis keeping the overall system functionality in mind.

Successfully integrating ergonomic principles requires the conduct of an unbiased trade study to determine the best equipment family for a particular application. This helps in preventing a conflict of interest between the ergonomics consultant and the client.

Successfully integrating ergonomics requires interpersonal skills, an understanding of the job environment, performing the proper analyses, and providing an unbiased solution.

REFERENCES

Chaffin, D.B., "Human Strength Capability and Low Back Pain," *J. Occup. Med., 16* (4), 248-254 (1974).

Chaffin, D.B., G.D. Herrin and W.M. Keyserling, "Preemployment Strength Testing," *J. Occup. Med., 20* (6), 403-408 (1978).

Department of Defense, *Human Engineering Design Criteria for Military Systems, Equipment and Facilities,* Military Standard (MIL-STD)-1472D, 1989.

Eastman Kodak, *"Ergonomics Design for People at Work,"* Volume 1. New York: Van Nostrand Reinhold, 1983.

Edholm, O.G., and Murrell, K.F.H., *"The Ergonomics Research Society, A History 1949-1970,"*: London: Ergonomics Research Society, 1973.

Gilmore, W.E., *"Human Engineering Guidelines for the Evaluation and Assessment of Video Display Units,"* Idaho National Engineering Laboratory, Prepared for the U.S. Nuclear Regulatory Commission, July 1985, pp. 449-479.

Godnig, Edward C. *Tips for Terminal Vision, Occupational Hazards,* August 1992, pp 41-43.

Keyserling, W.M., D.G. Herrin and D.B. Chaffin,"Isometric Strength as a Means of Controlling Medical Incident on Strenuous Jobs," *J. Occup. Med.,* 22(5), 332-336 (1980).

Miner, M.G. and J.G. Miner, *Employee Selection Within the Law*, Bureau of National Affairs, Washington, D.C., 1978.

Snook, S.H., The Design of Manual Lifting Tasks, *Ergonomics*, Vol. 21, No. 12, pp 963-985.

Van Overbeek, Thomas T., "*CRT Display Quality and Productivity*," Cornerstone Technology, March 15, 1991.

VDT News, *RSI Litigation Keeps on Growing*, July/August 1992, pp 1-10.

CASE STUDY QUESTIONS

1. According to this case study, how extensive is the ergonomic problem? Does ergonomics pay? Why?

2. Identify the social and legal issues associated with ergonomics?

3. Why was this case study unique? What role did the consultant play in this project? Is this a useful approach for an "in-house" safety professional? Explain.

4. Identify the problems associated with the original architectural design.

5. Identify the modifications made to improve the ergonomics of this office complex.

6. List the principles required to successfully integrate ergonomics into a workplace as presented by this case study author.

CHAPTER 2.8

SERVICE INDUSTRY CASE STUDIES

The case studies found in this section examine ergonomic issues in the service industry. They include materials handling problems, workplace environmental difficulties, and hand tool deficiencies. With the service industry experiencing the greatest growth in the United States today, ergonomic issues will gain importance in this industrial sector.

One of the case studies presented in this section lists the use of back belts as one possible solution to a material handling problem. The author of that case study, however, explains that research does not support the use of back belts as an effective hazard prevention method. This author wishes to reinforce the position in opposition to the use of back belts. Based upon personal experience and results presented in the literature, there are more effective alternatives for addressing back injury problems.

After reviewing the following case studies, identify the problems of concern in each one, list the strategies employed to address those problems, and develop a list of actions that you would take to improve upon the recommendations presented. In addition, consider reviewing the literature on the use of back belts in the occupational environment. List the advantages and disadvantages associated with the use of back belts. Determine your position on this controversial subject.

MANUAL MATERIALS HANDLING IN THE FROZEN FOOD DELIVERY INDUSTRY: DEPALLETIZING

Neil M. Brown

INTRODUCTION

This case study focuses upon depalletizing frozen food containers at a food distribution company. A food delivery company with two satellite depots was the organization of concern in this case study. It had an institutional division which serviced food items to finer restaurants and country clubs, and a home service division which delivered and stocked high quality frozen food items to individuals' family freezers. This company employed about 450 people. Since it was privately held, exact sales figures were not published. However, the two divisions had a total of approximately $75 million in sales each year. In addition to other functions, the satellite depots received shipments of fresh and frozen food from the processing plant, stored certain frozen foods in its large commercial freezer, and was a distribution hub of restaurant and home service products. Four to six drivers performed the activity of interest in this case study.

Arriving in a tractor trailer truck originating from the corporate main facility, containers of frozen food were stacked seven tiers high on a five inch high wooden pallet (Refer to Figure 2.8-2). The containers were made of molded, high impact plastic with good side handholds, and were 21 inches long by 15 inches wide by 12 inches high. Loaded weight was between 15 and 75 pounds with most containers averaging about 40 pounds. The tractor trailer was backed to a level loading dock where an electric pallet jack, steered by hand, was used to unload the pallets from the tractor trailer to the loading dock. From the loading dock, the pallets were either moved to a commercial walk-in freezer inside the warehouse, or left on the dock where the individual containers were immediately loaded into the local route delivery trucks. When the pallets were loaded into the warehouse freezer, the seventh tier was removed by hand because the door of the freezer only accommodated a six-tiered pallet.

The beds of most local route delivery trucks were between 16 and 18 feet long. The inside dimension of the truck bed was 92 inches wide and consisted of a rear door with a center aisle width of 28 inches and storage racks 32 inches wide on each side of the aisle.

Every morning four or five local route delivery trucks were loaded with approximately 100 of these containers of frozen food. The containers of frozen food were stacked onto pallets usually stored in the warehouse walk-in freezer. The local route delivery trucks were backed up to the loading dock. Pallets of random ordered containers were then stacked at the rear door of the local route delivery trucks using the electric pallet jack. Since the rear door of the local route delivery trucks was 28 inches wide, loading of each local route delivery truck was performed by hand. It was usually a two-driver operation.

Loading was done on a "buddy system" where the "depalletizing driver" searched for, removed and stacked containers. He then carried each teir of containers to the rear of the delivery truck that was to be loaded. At the same time, the driver of the truck was inside the back of his own truck loading the containers in a specific sequential order that would facilitate deliveries. Once one truck was loaded, the drivers switched roles and loaded the other local route delivery truck in the same manner. Each driver loaded the inside of his own truck with other drivers transporting containers to the end of the loading dock.

STATEMENT OF THE PROBLEM

Of the many ergonomic opportunities that could have been investigated in this procedure, this study was limited to the depalletizing of frozen food containers. Specifically, this study investigated the depalletizing person's ergonomic risk of developing lifting related low back pain while performing overhead reaching, lifting, and removing of upper tier containers and bending over to remove and lift lower tier containers. The 1991 NIOSH lifting equation was applied to the tier one and six analysis.

RESULTS

The following was observed for this materials handling task:

- Container handhold was lifted to 43 inches (elbow height of 50th percentile).
- Container's weight was assigned at 40 pounds average.
- Tier #1 handhold was 17 inches above the floor.
- Tier #6 handhold was 77 inches above the floor.

	LC	HM	VM	DM	AM	FM	CM
Tier #1	51	18"	17"	26"	30°	5/m	Good
Table		.56	.91	.89	.90	.60	1.0

RWL   = LC X HM X VM X DM X AM X FM X CM
       = 51 X .56 X .91 X .89 X .90 X .60 X 1.0
       = 12.49 LBS.

Lifting index   = Load Weight / Recommended Weight Limit
       LI   = 40 LBS. /12.49 LBS.
                 = 3.20

	LC	HM	VM	DM	AM	FM	CM
Tier #6	51	18"	77" (70)	34"	30°	5/m	Good
Table		.56	.70	.87	.90	.60	1.0

RWL   = LC X HM X VM X DM X AM X FM X CM
       = 51 X .56 X .70 X .87 X .90 X .60 X 1.0
       = 9.39 LBS.

Lifting index   = Load Weight / Recommended Weight Limit
       LI   = 40 LBS. /9.39 LBS.
                 = 4.25

RECOMMENDATIONS

Options that could be considered to eliminate this problem include:

1. Redesign the local route delivery truck, pallets and loading procedure. Redesigned pallets would be loaded at the main plant for sequential off-loading at the point of delivery. The redesigned pallets would be loaded directly onto local route delivery trucks by forklift. The local route delivery truck driver's first contact with the container would be off-loading at the point of delivery.

This would not be a viable option. All of the trucks would have to be redesigned and fabricated. A new pallet would probably have to be custom designed and custom fabricated. The forklifts used in this operation would probably have to be custom designed and fabricated in order to maneuver within the 92 inch (unloaded) width of the truck bed. This is not a reasonable option.

2. Use a multiple pallet stack. Instead of stacking containers seven tiers high on one pallet, stack a three tier pallet on top of a four tier pallet and use a forklift truck to remove the upper pallet.

The drawback to this option is that it would require the purchase and related expenses of a forklift truck which would have limited use at the facility. Additionally, maneuvering the forklift inside the warehouse freezer would be tight and risky. Forklift damage to the inside of the freezer would be expensive to repair and it could potentially cause the defrosting loss of the entire contents. This is not a reasonable option.

3. Use a conveyer. A battery powered, 10-foot long, narrow conveyor could be used so the exit end fits inside the truck's 28" wide rear door. This tiltable electric conveyer should be mounted on a chassis with wheels so the entire unit could be moved by the depalletizing person. The mobility and angle adjustability would permit the conveyer to be positioned so the feeding end would be at the pallet in the most comfortable positions possible. The exit end would be directed inside the local route delivery truck, height adjusted so the truck's driver could lift the containers comfortably. The pace of container feed/removal would be determined by the two operators, thus removing ergonomic metabolic concerns.

This appears to be a feasible option. It would additionally solve other ergonomic problems not addressed in this study such as carrying containers to the truck's rear door as well as spinal axial rotation and upper body load stress as the container is lowered through the 28" door opening. The cost would be "reasonable." This would be a one-time capital expenditure which could be easily maintained and costing less than the total medical expenses of only one cumulative trauma disorder.

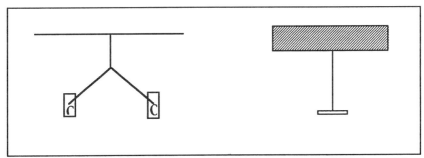

Figure 2.8-1: Examples of an adjustable electric conveyer (left) and portable trestle (right).

4. Use a portable trestle. A trestle is a support used for maneuvering loads on the point of balance to readjust the grip or carrying posture.

Figure 2.8-2: View of driver removing a container from the seventh tier.

Figure 2.8-3: View of driver placing container in delivery truck. Note how all containers are on pallets.

This would also appear to be a feasible and cost effective option. A portable, adjustable trestle with a compressive top surface would be used when depalletizing higher tiers of heavy containers. The compressive surface, instead of the depalletizer's musculoskeletal system, would cushion the fall of the container.

5. The main facility should load the pallets six tiers high by customer order/pallet unloading sequence. The seventh tier could be lighter containers since they would have to be removed for freezer entrance.

This option would eliminate unnecessary handling of every container on top of the tier since the local route delivery truck must be loaded in a delivery-stop sequence anyway. This option is better than no action and would require minimal planning, detail, or cost.

6. The main facility should load the pallets so lighter containers are on top and bottom tiers and heavier containers are stacked in the middle tiers.

This option is better than no action and would require minimal planning, detail and cost.

CONCLUSION

This task was ergonomically hazardous to all persons engaged in this activity considering the lifting index (LI) of tiers one and six was over 3.0 for the lifts alone, and did not include ergonomic risks associated with container carrying, spinal axial rotation under load, load extension and load pushing. Understanding cumulative trauma disorders as microtraumas that cumulative over time, the prudent preventive, as well as the ethical imperative is to redesign the job. The corrective options presented in this case study (the conveyer presented in Option #3 in combination with Options #5, or #6) would have had the greatest impact for the elimination of the ergonomic hazards. They would be minimally intrusive from a procedures perspective while remaining a reasonable "financial" option.

REFERENCES

How The Revised NIOSH Lifting Equation Works
...://tucker.mech.utah.edu/Pub/Tool/Rle/rlehow.html
MANUAL HANDLING - Risk Assessment
...ww.wt.com.au/safetyline/codes/manhan/man-ass.htm
Walk-Through Checklist For Manual Material Handling Risk Analysis
...//tucker.mech.utah.edu/Pub/Tool/Check/mmhopts.html
When To Use The Revised NIOSH Lifting Equation
...//tucker.mech.utah.edu/Pub/Tool/Rle/rlewhen.html

CASE STUDY QUESTIONS

1. What is the primary ergonomic issue of concern in this case study?

2. What was the problem analyzed? What tools and techniques should be used to analyze this type of ergonomic problem?

3. Calculate the RWL if only the tier 5 level was of concern in this case study.

4. If the containers were 25 pounds, what would be the lifting index in this case study?

5. What recommendation made by the author is the strategy that you would employ?

6. What additional strategies could be considered for this problem?

MANUAL MATERIALS HANDLING: LOADING OF A FREEZER IN THE FROZEN FOOD DELIVERY INDUSTRY

Neil M. Brown

INTRODUCTION

A food delivery company with two satellite depots was the organization of concern in this case study. It had an institutional division which serviced food items to finer restaurants and country clubs, and a home service division which delivered and stocked high quality frozen food items to individuals' family freezers. This company employed about 450 people. Since it was privately held, exact sales figures were not published. However, the two divisions had a total of approximately $75 million in sales each year. In addition to other functions, the satellite depots received shipments of fresh and frozen food from the processing plant, stored certain frozen foods in its large commercial walk-in freezer, and was a distribution hub of restaurant and home service products. The topic of interest is the loading of frozen food products into a residential freezer by a frozen food delivery company.

A local delivery driver's truck is loaded with containers of frozen food. The truck's freezer bed is about 18 feet long. It contains about 10 to 12 deliveries with each delivery consisting of 7 to 11 containers of frozen food. The frozen food items inside the containers ranged from a 10 ounce box of vegetables to a 7.5 pound ham to a fragile gourmet dessert cake. Most of the items were meats which weigh between 1 and 4 pounds per item. The majority of the items are irregularly shaped. The number of items to be placed into the customer's freezer range from about 65 to 175 per delivery. The containers ranged in weight from 15 to 75 pounds with most containers weighing about 40 pounds. After removing the containers from the truck's bed they were stacked four to five high in the customer's driveway. The driver then used a handtruck to move the stacks of containers to the family's freezer. After the stacks were moved to the freezer, the delivery driver opened the door of the family's freezer, opened the containers of frozen food and, piece by piece, the freezer was loaded with the food items from inside the containers.

STATEMENT OF THE PROBLEM

This study was limited to the loading of a residental freezer with items from the containers. Specifically, this study investigated the delivery person's ergonomic back injury risk by the process of repeated bending, stretching, and twisting to grasp irregularly shaped frozen food items from the delivery containers (Refer to Figure 2.8-4).

The 1991 NIOSH Lifting Equation was applied to two lifting positions observed. The first position was food items removed from containers from the floor level to a midpoint in the freezer; the second position was from a container stacked on top of three other containers to the same midpoint destination in the freezer as the first observation. The Lifting Index and individual multipliers of the Revised NIOSH Lifting Equation will be addressed.

RESULTS

The two positions studied were containers on the floor (f) and a container trestled (t) on top of three other containers in front of the freezer. Both positions required the delivery person to lift objects from the container and place those objects into the freezer. The containers of frozen food placed on the floor were spread out, thus further from the freezer than containers stacked or "trestled," requiring a greater twist of the delivery person's torso.

	f	t
LC	51 lbs.	51 lbs.
HM	25"/.40	15"/.67
VM	0"/.78	40"/.93
DM	40"/.87	0"/1.0
AM	90 deg./.71	30 deg./.90
FM	.45	.45
CM	.90	.90

Each item lifted from the container was a different size, shape, and weight from the other items lifted. Each item was at a different vertical location at the lift's origin as per placement within the container and had a different surface moisture content affecting coupling. Each item also had a different destination inside the freezer being stocked resulting in different vertical distances from the floor as well as different horizontal placement inside the freezer. For ease of comparison, certain dimensions were given the same measurements with the dimensions presented in the table above reflecting that compromise. The frequency multiplier (FM) was listed at 10 lifts per minute for a duration of less than one hour; vertical (V) location of the hands at the origin of the lift, $V < 30$ or $V \geq 30$, did not change the tabular listing. The coupling multiplier (CM) was listed as "poor" because cotton gloves with no provision for an anti-slip surfaces were used to grip

Figure 2.8-4: Two views of container product being transferred to a
residential freezer.

smooth, irregularly shaped, moisture/frost covered items. Two pounds was assigned as the weight of the item lifted.

RWL   = LC x HM x VM x DM x AM x FM x CM
RWL(f) = 51 lbs. x .40 x .78 x .87 x .71 x .45 x .90
       = 3.98 lbs.

RWL(t) = 51 lbs. x .67 x .93 x 1.0 x .90 x .45 x .90
       = 11.58 lbs.

Lifting Index

The lifting index (LI) is a mathematical treatment which provides a relative estimate of the physical stress associated with a manual lifting job.

Lifting Index = Load Weight / Recommended Weight Limit

LI = LW/RWL

LI(f) = 2 lbs./3.98 lbs.
      = .50

LI(t) = 2 lbs./11.58 lbs.
      = .17

CONCLUSIONS

Since the lifting index of each lifting task studied is below 1.0, the Revised NIOSH Lifting Equation concludes neither task poses "an increased risk of lifting-related low back pain for some fraction of the workforce." However, some of the frozen food items lifted exceed 4 pounds. With that figure used in the lifting index for the floor lift, the lifting index is >1.0. This condition does "pose an increased risk for lifting-related low back pain for some fraction of the workforce. NIOSH considers that the goal should be to design all lifting jobs to achieve a LI of 1.0 or less."

Additionally, the individual multipliers can be used to identify specific job related problems:

- When the horizontal multiplier is less than 1.0, the load should be brought closer to the worker. The trestle lift is closer to the worker thus a safer procedure than the floor lift.

- When the vertical multiplier is less than 1.0 the origin of the lift should be raised. Lifting from the floor should be avoided.
- When the distance multiplier is less than 1.0 the vertical distance between the origin and destination of the lift should be reduced as much as possible. The goal of the trestle lift is to reduce the vertical distance between the origin and destination of the lift.
- If the asymmetric multiplier is less than 1.0 the origin and destination of the lift should be brought closer to reduce the angle of twist. Since the trestle lift is closer to the destination, it reduces torso twist.
- When the frequency multiplier is less than 1.0 the frequency, duration or recovery period should be altered. I do not believe this is a reasonable or practical factor to alter in this job.
- If the coupling multiplier is less than 1.0 the ability to hold the items lifted should be improved.

RECOMMENDATIONS

Some of the following suggestions may not be practical, may cause more harm than good or may be controversial:

I.   Provide a Barnett Back Strap which actually requires proper lifting technique by preventing the back from bending to the dangerous 70 degree angle. (Best's Safety Directory, 1995)

II.   Provide a Back Alert, an electronic device worn on the belt which emits an audible signal when the wearer bends to the dangerous 70 degree angle. This is a behavior modification tool as it reminds and reinforces proper bending techniques with its audible cues. (Best's Safety Directory, 1995)

III.   Train the workers and equip them with back support belts (Material Handling Engineering, 1994). However, back belts can do more harm than good (Business Insurance, 1995). Controlled studies have found belts to be of little or no value in back injury prevention (Professional Safety, 1994). NIOSH does not recommend the use of back belts among uninjured workers and does not consider back belts to be personal protective equipment (Occupational Hazards, 1994). Also, although the concept of training workers to lift safely appears valid, the results have been poor. Uninjured workers are not easily motivated, training quality varies, compliance is inconsistent and "safe" lifting is not natural (Professional Safety, 1994). Imposed procedures against the will of the worker can not be enforced when the worker is alone.

IV.   Training Procedure. Train the drivers to use the trestle concept whenever possible. Train the drivers of the increased risk of back injury from lifting items off the floor. If the driver would simply use a full, or empty closed, container as a trestle one, two or three containers high, the vertical location of the hands to the floor would be improved, the distance traveled from the origin to the destination of the lift would be improved and there would be less need to bend, reach and lift items from containers spread out on the floor.

Training procedure (as recommended by Mr. J. Patterson):

A.   Think before acting; fill the freezer mentally first. From the computer printout of the items in that customer's order, mentally fill the freezer from looking at the list. An experienced driver knows the shape, size, and probable destination of most items. Look at the list, look at the inside of the freezer and fill it mentally first.

B.   Estimate from the weight and feel of the container what kinds of items are probably inside.

C.   Stack the containers per best estimate with the items going to the top shelf in the highest container, middle shelves to the in-the-middle stacked containers and bottom shelved things on lower containers. Although not completely accurate, this step reduces much of the distance items need to be lifted.

D.   At the lower shelves, get onto your knees; pull the container closer to you if necessary. This keeps the back straighter, you don't have to bend, reach and twist as much. (Use a couple of layers of cardboard to cushion your knees from the floor. This also keeps your pants from getting dirty.)

E.   Buy your own gloves. It is recommended that gloves with a rubberized flexible surface be used.

It appears that the above-mentioned training option would be the most cost effective option to consider. By addressing the fundamental issue of job process design instead of adding personal protective equipment, many of the ergonomic hazards have been eliminated. Not all the hazards have been eliminated, but this would improve the task and increase productivity (Progressive Grocer, 1994). Reducing angular movement and improving the load's control and coupling can reduce the risks. (Occupational Health & Safety, 1995)

This procedure is cost effective; it costs very little to implement. It would increase productivity and decrease time spent on the task because of better organization. It reduces the amount of movement necessary and the fatigue associated with repeated bending, lifting, and twisting. The delivery person presents a more organized, thus more professional, presentation to the customers who watches their freezers being stocked. The potential expense of future medical claims for lifting related injury should also be reduced.

Additionally, the individual multipliers listed in the Conclusions section are improved by this method. The horizontal location of the hands at the beginning of the lift are closer to the saggiteal plane. The vertical locations of the items are closer to the vertical destinations. Since the stacked containers are closer to the opened freezer, there is less spinal axial rotation. The improved gloves bought helps increase the coupling modifier from "poor" to "good." This procedure is inclusive of and supersedes option #4. It is further recommended that the company supply the drivers with knee pads that are comfortable, easily used and easily carried.

REFERENCES

Calmbacher, C. W., (1995, February). Software Developers Help Analyze Task Hazards, NIOSH Lift Calculations. *Occupational Health & Safety*, 64, 37.

Chandler, R. L. (Ed.), (1995 Edition). *Best's Safety Directory*, 1995.

DeVor, R. E., Chang, T., Sutherland, J. W. (1992). *Statistical Quality Design And Control: Contemporary Concepts And Methods*. New York: Macmillan Publishing Company.

Employers Insurance of Wausau. (1994). *Material Handling Your Questions Answered* (Satellite Conference Booklet).

Fletcher, M. (1993, July 25). Take Back The Belts? Maybe. *Business Insurance, 28*, 2, 17.

Garry, M. (1994, January). It's Not Weird Science. *Progressive Grocer, 73*, 67 - 74.

How The Revised NIOSH Lifting Equation Works. University of Utah Research Foundation.
...://tucker.mech.utah.edu/Pub/Tool/Rle/rlehow.html

It Pays To Know Ergonomics Equipment (1994, April). *Material Handling Engineering, 49,* 58 - 63.

Kroemer, K. H. E., (1980, February). Back Injuries Can Be Avoided. *National Safety News,* (reprinted in *Material Handling Your Questions Answered* by Wausau Insurance Co., 37 - 43).

LaBar, G. (1994, September). NIOSH Challenges Back Belt Use. *Occupational Hazards, 56,* 61 - 64.

Mahone, D. B. (1994, July). Manual Materials Handling: Stop Guessing And Design. *Professional Safety, 39,* 16 - 19.

Maier, N. R. F. (1970). *Problem Solving And Creativity In Individuals And Groups.* Belmont, CA: Brooks/Cole Publishing Company.

Manual Handling - Risk Assessment (1995, 1 October). WorkSafe Western Australia Commission, Code of Practice.
..ww.wt.com.au/saftyline/codes/manhan/man_ass.htm

Walk-Through Checklist For Manual Material Handling Risk Analysis. University of Utah Research Foundation.
...tucker.mech.utah.edu/Pub/Tool/Check/mmhopts.html

Walk-Through Checklist For Upper Extremity Cumulative Disorders (1994). University of Utah Research Foundation.
...//tucker.mech.utah.edu/Pub/Tool/Check/uectd.html

CASE STUDY QUESTIONS

1. What was the ergonomic problem of concern in this case study?

2. What factors made this ergonomic problem unique and some what difficult to address?

3. Was the load being lifted a serious ergonomic problem? What factors brought you to this decision?

4. List the strong and weak recommendations in this case study.

5. What would you recommend to address this ergonomic problem?

6. What is your opinion on the use of back belts? What are the advantages and disadvantages of the use of back belts? Would you recommend their use? Why?

AN ENVIRONMENTAL-ERGONOMIC CASE STUDY AT A PARCEL DELIVERY FACILITY

Allen B. Tillett

INTRODUCTION

On Monday, June 19th, a memo was sent to the local terminal of a parcel delivery company. The memo came from the corporate office claims department regarding the terminal's worker's compensation expenses over the previous fiscal year. Headquarters told local management to investigate the problem and reduce accidents at any cost. Further, they were to supply the reason for the problem and a feasible solution to eliminate the cause.

CASE STUDY

A team of two shift coordinators and the safety manager was organized to investigate the problem. A review of the OSHA 200 logs revealed that the third shift had numerous accidents that appeared to be ergonomically related. Many accidents appeared to have occurred on the morning shift due to the early start up time of 3:30 A.M. Other accidents occurred because employees were reported to have been in a hurry to complete the work task as quickly as possible and thus made avoidable mistakes or misjudgments that resulted in injuries.

The first step of this ergonomic assessment was to interview the employees to determine why there were so many accidents occurring during the previous period. Some employees had no opinions and others felt that injuries were expected with the kind of work they were employed to perform (package handling from a commuter van to a conveyer belt to a linehaul trailer). However, one employee suggested that recent accidents occurred when employees stepped from the concrete pad under the conveyer belt up to the commuter van. As a result of this comment, the investigation team reevaluated the OSHA 200 log and the supporting accident report documents. It was found that approximately 87.3% of all accidents occurred on or near the commuter van.

This employee observation appeared to be the best explanation for the recent rash of accidents. These findings were sent to corporate headquarters while, at the same time, the safety committee began to take suggestions on how to resolve this problem. Suggestions were received from both the employees and upper management at the terminal. Before the committee had an opportunity to compile these responses to solve this problem, corporate headquarters sent another memo stating that they had resolved the problem. Headquarters concluded that this terminal was one of the newest terminals and it still had a gravel parking lot. The gravel parking lot was frequently exposed to rain which made the surface uneven with the loading dock. Further, the traffic associated with the vans entering and departing the terminal displaced the gravel. This steady stream of traffic created an uneven surface and the van bumpers were therefore not at the same level as the loading dock. It seemed as if this problem was solved.

It was just a matter of time before the parking lot was paved. However, one member of the safety/ergonomics team felt that the corporate headquarters suggestion was less than adequate. He felt that the paving of the driveway and parking lot would not completely solve this ergonomics problem. He took measurements from the van floor to the ground and determined that each van was different. It was suggested that measurements from the bumper of the van to a level surface be taken since the bumper was used for a step to gain entry into the van for loading and unloading activities. When these measurements were taken the conclusion was reached by the ergonomics committee that paving the driveway and parking lot would not solve the problem.

The team met once again, presented the committees findings to both employees and management, and asked for recommendations based upon this new evidence. The team agreed that the measurements were correct and that corporate headquarters should be notified to hold off on paving the parking lot. However, there was resistance from upper level management at the terminal because they wanted to have a paved parking lot to reduce the dust being thrown onto their personal vehicles. They suggested that it would also create a more professional look to the terminal. However, they were convinced that the paving of the parking lot was forthcoming as the funds were to be provided in the coming year and that the primary objective of the committee and the terminal was to reduce the frequency of accidents.

RECOMMENDATION

Upper level management wanted the parking lot paved. However, one of the managers presented an alternative solution to this problem. He recommended that a ramp be ordered or constructed and placed between the incoming vans and the dock area floor. This idea came about as a result of that manager witnessing this application at another terminal. This other terminal appeared to have the same problem. It was later determined that a

Figure 2.8-5: A side view of the load dock (left side of photograph) and the bumper of the delivery vehicle (right side of photograph). Notice the difference in height that contributed to the hazardous environment.

manufactured ramp would serve the same purpose at a minimal cost; it had solved the other terminals accident problem.

Since installing the ramp the number of injuries for the terminal is below the previous year's average. The committee viewed this result as their first success.

CASE STUDY QUESTIONS

1. Who were the members that made up the ergonomics team? Is this committee representative of all the key players at this organization? Is it important to have all key players represented? Why?

2. What were some of the comments obtained during the employee interviews? Are these typical responses? How would you reply to these comments?

3. What statistic directed the safety committees attention to the location of concern at this facility? Where were these statistics recorded?

4. Why did the safety/ergonomic committee feel that the corporate response was less than adequate?

5. What solutions recommended by the committee do you think would best solve the ergonomic problem?

6. What would you recommend to improve the facility's ability to control the ergonomic hazards identified?

ERGONOMIC TOOL DESIGN HAZARDS
IN THE RESTAURANT INDUSTRY

James E. Krouse

INTRODUCTION

This case study represents finding ergonomic solutions for shoulder, elbow and wrist strains associated with restaurant work primarily among female employees ages 18-23 years.

The organization of interest in this case study was a chain of pizza restaurants located in the Midwest. At the time of this study, there were approximately 20 locations in the Midwest. Employees in this chain were approximately 65 percent female with their average tenure being three years or less. The employee workshift averaged 8 hours or less, with employee duties ranging from waitress and kitchen help to individual unit managers and assistant managers. Most employees in this organization worked approximately 8 hours per shift.

The organization had initiated at the time of this study a minimal amount of safety training and virtually no studies or evaluations had been completed in the ergonomic area. Employees were hired after an application had been made, given a minimal amount of instruction that was required by the home office and allowed to begin their duties. There was no known training in regard to the stressors of the job for new or existing employees. There was also a minimal amount of data collected, beyond insurance company loss runs.

STATEMENT OF PROBLEM

As this particular chain of restaurants grew, the work activities required at some of the more high volume units was greatly increased. As a result, workers' hours were often expanded and thus, the physical demands of the job increased. Approximately one-third of the total number of units had sales that were far in excess of the national average for this national franchise.

Job demands had increased to the point that unit managers and area managers were starting to see an increase in carpal tunnel syndrome claims, tendonitis, elbow and joint problems associated with one phase of this operation. It was felt by management that the configuration and kitchen activities associated with removing and storage of pizzas prior to being delivered to customers was creating the bulk of these stressors. The employees that were suffering these injuries were 100 percent female employees. Although male employees did perform some of the kitchen work associated with pizza baking, most of the routine activities were accomplished with female staff.

CASE STUDY

Employees were required to remove pizza and specialty sandwiches from a double oven conveyor system and place them in a warming mode on the upper portion of the ovens that were in the kitchen. These types of ovens are common in this industry, as they allow for quick food preparation with minimal amount of special requirements. The double tiered ovens were approximately 5 feet 8 inches high by 5 feet in depth. The ovens allow individual food items that have been prepared to be placed on a slow moving conveyor on one end of the oven and moved on another. They not only allow for relatively quick food preparation, but provide quick access of hot food to the customer.

During peak hours of approximately 5:00 p.m. to 10:00 p.m., female kitchen help would have to remove recently cooked food and place it on the top portion of the oven for temporary warm storage. It not only allowed for additional space for preparation of new food items, but provide a limited amount of warming and retention area prior to serving items to the customer. Managers felt that this process in some way may be the culprit in the ever increasing physical problems associated with their female staff.

A safety consultant was contacted by the chain to make an evaluation of the problem and to offer solutions for risk improvement. The consultant found the following areas to be of concern during initial videotaping and physical observation of a random number of locations. Those items of concern that were noted were:

1. The kitchen areas where food was being prepared had a very limited amount of work area for the volume of sales of each restaurant.
2. The floor surfaces consisted of quarry tile with a smooth surface that was often greasy and slippery as a result of food items dropped on the floor.
3. Employees were using a standard 10-inch channel lock pliers in all locations, to lift pizzas and other food items from the double conveyor ovens (see Figure 2.8-6). These

pliers were purchased from local hardware stores and were used instead of hot pad or asbestos gloves which, during an initial investigation, were found to wear out quickly.

4. Employees who retrieved the pans from the conveyor ovens with the channel lock pliers would frequently have to place heavy food products on the top of the double oven. This was found to be beyond the normal range of motion. This practice placed a great deal of stress on the upper shoulder, elbow and wrist area of female employees. The average height of employees in the chains sampled were less than 5 feet 5 inches tall. The food items that were retrieved from the double conveyor oven and placed on the tops of the unit had a weight varying from approximately one pound to nearly seven and a half pounds for a deep dish pizza. Pliers utilized would grip a very small portion of the aluminum baking.

5. During analysis of insurance records, the consultant found that a number of liability claims had occurred as a result of the locking nut on the pliers coming loose and being baked into pizza or food product. This rather precarious situation often ended in broken teeth with customers. Employees up to this point had been warned to watch the locking nut on all mechanical tool devices including circular pizza cutters.

6. There were no stools that were utilized to provide better access to the tops of cooking ovens for storage purposes.

7. The consultant found during random interviews with female employees that as many as 1/5 of all employees who were completing this task, were suffering injuries from mild discomfort to carpal tunnel syndrome.

RECOMMENDATIONS AND CONTROLS

During his analysis, the safety consultant determined that increasing the size of work areas and physically changing the configuration of ovens would be difficult and not cost effective. It would have required extensive remodeling and purchasing of new, single level conveyor ovens that were larger and more user friendly. As a result, training and tool design were focused on as primary solutions.

The following solutions were presented to the management of this chain:

1. The present tool devices that were utilized were found to be inadequate and dangerous as used in this application. It not only created a liability hazard, but because they were not ergonomically designed, put a great deal of stress on employee's hand, wrist and arm areas when elevating product above shoulder height for storage purposes. It was determined that new prototype tools should be constructed and tested at various locations for integration throughout the chain. An Indiana-based firm was contacted to accomplish the task.

    An ergonomically designed tool that curved into the palm area of the employee was constructed to provide less stress to the employees than the existing tool. The tool was constructed of lighter aluminum, had a permanently fixed nut area and the head was designed specifically to accommodate pizza pans and specialty foods. The tool also maintained the wrist in a neutral position to avoid flexion deviation. It was also felt with the new tool devices, that employee theft would be reduced. This was because of the specialty design of the product and its usefulness beyond the kitchen (see Figure 2.8-7).

2. To implement change in the organization was difficult. Management originally integrated the new tools directly along side the existing pliers that were present at each location. Employees in this respect were often reluctant to utilize the new tooling because of their familiarity with the existing pliers in each unit. The consultant felt to properly facilitate the change would require the removal of the old tools as soon as feasibly possible. Employees after the re-education process and instruction on their use and new design would accept them as viable alternatives.

Figure 2.8-6: Standard 10 inch channel lock pliers originally used by employees to remove product from ovens.

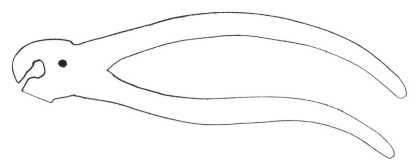

Figure 2.8-7: New ergonomically correct hand tool introduced to the pizza restaurant chain.

CASE STUDY QUESTIONS

1. What other areas that were mentioned in the case could be improved in the kitchen work station environment?

2. Why is training in this case a major element of concern and what types might be covered?

3. What should management's role be in the future with these issues?

4. New employee orientation is always important. Discuss your approach with the new tool.

5. Why are anthropometric issues very important with this case?

WORKPLACE EQUIPMENT AND FACILITY REDESIGN IN THE RESTAURANT INDUSTRY

James E. Krouse and R. Scott Lawson

INTRODUCTION

In 1995, an employee of a national pizza chain suffered a fractured neck as a result of performing maintenance duties as a working supervisor. The employee was stacking conference tables, as part of his routine job, when they slid out of position striking the employee. Injuries were in the upper extremity of his back and spinal area fracturing a vertebrae. This debilitating injury required the employee to spend three months in the hospital. It was not known if the employee would walk as a result of the injury sustained.

The 36-year old employee had been employed with this restaurant chain for nearly 20 years in a maintenance capacity. Although the employee was considered to be the manager in the maintenance department for this chain, he was considered a working supervisor. He was involved in all facets of maintenance, including the repair and maintenance of ovens, walk-in coolers, refrigeration units, dough mixers, compressors, landscaping, lighting, sprinklers, fencing, repair and replacement of water heaters and working on roof heating and cooling systems.

The injuries sustained by this employee placed him in a questionable situation for long-term employment with his employer. It was felt at the time of his injury that the employee would probably never return to the current position in which he was originally employed.

This case study took place at a 50-unit pizza franchise located in the Midwest. There were more than 300 individuals employed by this chain. A maintenance department and office staff were maintained to support existing business activities throughout the franchise. This particular chain had been in business approximately 25 years.

STATEMENT OF PROBLEM

The problem of concern in this case study was to retain an individual with physical limitations in a maintenance management position. The Management of this chain of pizza restaurants was determined to rehabilitate this employee. Their purpose was to place this individual in as close to his pre-injury work condition as possible without risking further injury. Management also wanted to assist in providing improved work methods and procedures. They felt this would not only enhance this employee's over all activities, but would also assist in reducing stresses to other employees in their growing business.

CASE STUDY

It was felt after review of existing medical data and information provided by physical therapists, that the employee, because of permanent nerve damage, would have major limitations in his work environment. The physical therapy unit treating this employee felt that an independent consulting firm utilizing ergonomic training and sound engineering practices could provide improved "safe" work methods.

The following work processes were assessed as part of an initial review by the consulting firm. These were:

- Identification of possible work objectives.
- Review of basic work skills and tolerances.
- Review of work habits and interpersonal skills with peers.
- Review of work-related capabilities, such as transportation, communication and self care.
- Review of ongoing work handling programs.
- Review of applicable work adjustment habits.
- Review of tasks and environmental adaptations.

The consulting firm was particularly concerned about the employee's permanent physical weaknesses. They were concerned about the employee's grip strength and mobility. In addition, the consultants felt that the employee would have to be restricted to lifting no more than 40 pounds. An initial concern to the neurosurgeons and physical therapists was the injured employee's ability to grip material and balance himself while climbing ladders and working on elevated platforms. It was estimated, by the consulting firm, that these tasks amounted to 5 to 10 percent of the injured employee's work schedule.

The consultants felt that material handling issues associated with heavy materials being lifted by this employee could be approached by using dollies, carts, hydraulic lifting devices and assistance from other employees. These issues were covered in detail in the report with both management and the employee. Of particular importance to both management and the employee was maintaining individual independence while performing his work and allowing him to continue to work at heights.

The remainder of this case study will deal with the issues of climbing, use of ladders and moving materials to elevated roof surfaces. These presented the biggest challenge to the consultants because of the unavailability of current products in the workplace. Most ladder and elevated roof work activities were accomplished usually by one individual. This made the challenge of providing sound recommendations for risk improvement important.

RECOMMENDATIONS AND CONTROLS

Current ladders that were used in this organization were ANSI Aluminum Type III Extension Ladders that were provided to employees and mounted on a standard full-size Ford Econoline Van. The approximate height of the van roof area and access points was 7 feet. Under normal circumstances, employees within this chain were required to stand on the back bumper system, unlatch the ladder and pull a singular 14-foot section backwards to dismount the ladder for placement. Due to the employees' limitations in regard to balance and lifting capabilities, this made this activity nearly impossible. Another consideration was to provide stable access of the employee to the roof and balancing tools and other materials.

The following recommendations were made:

1. Provide a ladder that has a weight of no more than 40 pounds., including fabricated hand rails to provide stability for the injured employee. Specific considerations were utilized in regard to current OSHA Standards with the design (see Figure 2.8-8).
2. Provide a mounted hydraulic dismounting device on the employee's van to assist in proper placement of the ladder from the roof system to chest height (see Figure 2.8-9). The cost for adapting similar hydraulic discharge systems for this task was approximately $5,000.
3. Provide a front loaded tool holding vest that could be counterbalanced. This would eliminate the dangers of utilizing a tool belt that provided instability to the employee because of permanent nerve damage and weakened lower body strength. Upper shoulder and body strength was considered adequate for this tool belt design. This design would also free the employee to grip the newly designed ladder for increased stability.
4. Provide a ladder building stand and mechanical device to hoist material to the roof where needed. This building stand and mechanical device would be attached to the ladder. It would also add stability to the unit

and would provide easy access to heavy material that could not be carried by the injured employee. The hoisting design is a double-handled unit that would be part of the building stand and would be cranked with both the left and right hand of the employee for stability (see Figure 2.8-10).

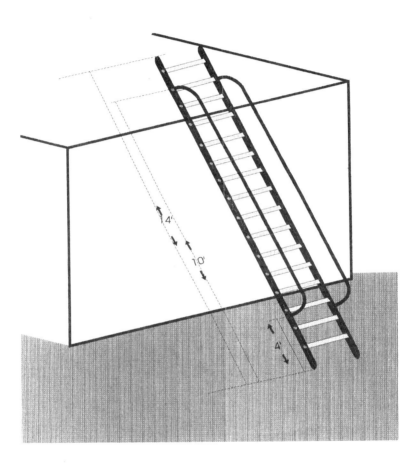

Figure 2.8-8: The handrails on the ladder pictured will have to be installed after purchasing. The handrails should be of light aluminum tubular construction. Ladder weight without the handrails is 23 pounds.

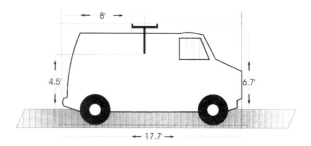

Figure 2.8-9: This figure denotes location of hydraulic unit on the van. When unit is actuated, the arm will lower to the side of the van at mid-chest height.

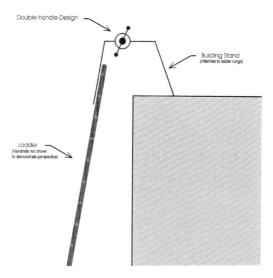

Figure 2.8-10: The current ladder crane used should be redesigned to incorporate a double-handle and building stand to alleviate the amount of force required to crank objects to the top of the building.

CASE STUDY QUESTIONS

1.   What ergonomic considerations should be investigated when dealing with individuals in the maintenance field? Give reasons and explanations for your answer(s).

2.   How would grip strength affect an individual's capacity to climb? Name and describe at least three different types of grips and determine which type of grip is the most ideal.

3.   What are a few design guidelines for handles?

4.   Name and describe at least three material or container characteristics and four task and environment characteristics that influence the suitability of manual handling.

5.   What risk factors are associated with poor cart design? Describe each risk factor.

CHAPTER 2.9

UTILITY INDUSTRY CASE STUDIES

The case study in this section examines ergonomic signal detection issues associated with the Three Mile Island (TMI) nuclear disaster. Displays and controls have served an important role in ergonomics for many years. The layout and design of controls in aircraft and military equipment is well documented. This particular problem is very interesting because most of the recommendations found in the literature was apparently not used in the control room at TMI.

After reviewing the following case study, identify the problems of concern associated with the displays found in the TMI control room. Find equipment that have displays and study the characteristics of those displays. Compare the characteristic that you find to those recommended in the ergonomics literature.

SIGNAL DETECTION — THREE MILE ISLAND ACCIDENT:
RIGHT STRING WRONG YO YO

Ron Pate

INTRODUCTION

I once took my Sears' model Weedwacker® to a Homelite® dealer for repairs. The repair man shrugged his shoulders and grinned, "Can't help ya son. It's the right string, but wrong Yo Yo." These were my thoughts as I read about the sequence of events that led to what arguably was the most bungled nuclear plant accident in history: Three Mile Island.

The nuclear power plant was owned by Metropolitan Edison (Met Ed), a subsidiary of General Public Utilities (GPU). GPU was the first utility to build a non-government subsidized nuclear power plant in the United States in 1963 (TMI Vol. I-2).

The power plant consisted of two 1300 megawatt nuclear reactors, designated Units One and Two. At the time of the accident, Unit One was shut down due to refueling, but was scheduled for a restart later in the day on March 28, 1979.

The plant was quite automated, requiring just four operators to run the entire graveyard shift. Many of the plants operators were young veterans who graduated from the Navy's Nuclear Control Program. They had limited experience with nuclear power plants like the units found at Three Mile Island. Most of their experience was on much smaller nuclear power plants working within the confines of a strong support military program.

At approximately 4:00 a.m. the sequence of events began to unfold. A pressure relief valve (PORV) designed to maintain a stable coolant pressure to Unit Two's reactor became stuck in the open position. This allowed coolant to start draining form the primary cooling system (PCS), causing pressure to drop and temperatures in the reactor to rise. As more and more coolant left the system, the top of the reactor core became exposed and a partial melt down of the reactor core began.

Two hours and eighteen minutes later, an operator from another shift finally deciphered the array and conflicting signals and took the appropriate corrective action. By determining that the PORV was stuck open, and shutting off a back-up valve, the operator had prevented a total core melt-down. Still, it would be another ten hours before anyone would turn back on the high pressure pumps to replenish the coolant that had been lost due to the open valve. All this time the reactor core was exposed and overheating. Other safety systems tried to correct the low coolant/overheating problem. Most of these systems were damaged in the process and became non-functional (TMI Vol. II 9-47).

To more fully understand how such a series of events could occur, it is important to understand the working environment, according to the investigators:

> "...the control room itself, for all the array of instrumentation has an improvised look; it would **not** be mistaken for the starship Enterprise. The control panels nearer at hand are arranged in an arc, effectively fencing off operators from annunciation panels that stand at the far side of the room, and display banks of alarm signal. An operator needing quick access to these more distant panels make a choice between an end run, and a vault. Repair tags hang from dozens of toggles, some of them obscuring the faces of the instrument" (TMI Vol. I-9).

Investigators noted that during the first two hours, over 100 of the 750 alarms present in the control room sounded. It was impossible for the control room operators to decipher all the information from the over 2000 indicator lights that had illuminated (TMI Vol. II-577).

STATEMENT OF THE PROBLEM

It was clear that fundamental mistakes were made with the design of the control room and its displays. It appeared that more emphasis was placed on getting information to the control room than in making sure that the information could be detected, identified, and discriminated (Gopher & Kimchi-423). Still, it is hard to believe that for almost 13 hours, technicians, engineers and even the plant's designers failed to grasp the situation and take full corrective action. Part of the reason lies in the fact that a fundamental axiom of display design had been violated: A display must represent properly the world that one wishes it to represent (Gopher and Kimchi-423). Three Mile Island accident investigators found that the PORV indicator light on the control panel did not indicate that the PORV valve was open as labeled. It only indicated that the pressure sensors were sending an electrical signal to the PORV solenoid to open. The PORV's opening and closing was strictly mechanical and was not displayed on the instrumentation on the control panel. For an operator to determine if the PORV was actually open, procedures required that the operator look at PORV discharge, and PORV drain tank temperature displays for an elevated temperature reading. The sequence of events that occurred are summarized as the following:

Once the operators noticed that the primary coolant system was losing pressure, their first reaction was to look at the PORV to see if it was open. The indicator light was off, correctly indicating no power to the solenoid, but this was misinterpreted as meaning the PORV valve was closed. Next, the operator checked the two temperature displays and noticed elevated temperatures. However, because of leakage problems with the PORV during the previous month, operators had become accustom to seeing elevated temperatures. These signals were entirely disregarded. The operators then incorrectly diagnosed the problem as too much water in the PCS system. Their reasoning was that too much water in the PCS must be over cooling the reactor, causing a reduced pressure and temperature within the PCS. They then shut down the emergency high pressure water pumps that had automatically tripped on to correct the loss of coolant. In the next two hours water continued to exit the primary cooling system through the stuck PORV. The reactor core became exposed resulting in a partial core melt-down (TMI Vol. 1 10-13).

A Special Inquiry Group asked that a human factors investigation be conducted to determine specific cause of this accident and to make recommendations for improvement. The recommendations made by this group were quite numerous and only a few, as they relate to enhancing information transmittal (signal detection), are listed below:

- The PORV system needed a sensor that directly sensed the mechanical position of the valve.
- Re-design the workstation to include placing instruments logically and in accordance with physical plant layout.
- Layout of instrumentation must be visible, accessible, properly labeled and marked, alignment of displays should be such that they enhance error detection. Work repair tags must not be left attached to control panels.
- Standardized color schemes, coding, and enhance contrast on lighted indicators to aid in message detection.
- Re-design meters, standardize scale marking, number progression, operating ranges must be coded; i.e., normal, high, low, danger.
- Certain systems need to record data for analysis not just display current data.

(TMI Vol. II 574-609)

I learned by taking my Weedwacker® to a Homelite® repair shop, that getting it there, did not necessarily mean it was going to get fixed. Three Mile Island taught the world that just because you can recieve information on a control panel, it does not necessarily mean it can be used to solve a problem. This is a situation where you have the right string, but the wrong yo yo!

REFERENCES

Nuclear Regulatory Commission. Three Mile Island: *A Report to the Commissioners and to the Public. Vol. I & Vol. II,* Washington D.C. April 5 1979.

Gopher D. & Kimchi R. *Engineering Psychology.* Annual Review of Psychology 40:431-55. 1989.

Meshkati N. *Ergonomics of large-scale Technological Systems.* Impact of Science on Society, no 165, 87-97.

CASE STUDY QUESTIONS

1. List the ergonomic hazards associated with this case study.

2. What operator factors may have contributed to the Three Mile Island incident?

3. What equipment and display problems contributed to this incident?

4. List what you consider to be the most important ergonomic recommendations made by the accident investigation team.

5. What would you look for regarding equipment displays and controls for your facility that would eliminate similar problems to those experienced at TMI?

Index

Boldface [#] denotes section

A

Absenteeism, **[1]**7, 27
Air contaminants, **[1]**11
Anthropometrics
  chemical industry, **[2.2]**2–4
  principles of, **[1]**9–11
Arm injury, **[1]**18
Assembly, see Manufacturing and assembly
Attitudes of workers, **[1]**10
Automobile radiator parts manufacturing
       facility, **[2.5]**17–19

B

Back Alert, **[2.8]**13
Back injury, **[1]**19–22
  cumulative trauma, **[1]**17
  in electronics industry, **[2.3]**2–4
  in hospital, **[2.4]**2–4
Back support equipment, **[1]**22, **[2.8]**13
Behavioral engineering approach, in manufac-
       turing, **[2.5]**2–12
  baseline analysis, **[2.5]**7
  intervention, **[2.5]**8
  safe performance observation data,
       [2.5]9–10
  setting, **[2.5]**6–7
Behavioral factors, **[1]**9–10
Behavioral sampling strategies, **[1]**10–11
Benefits of ergonomics programs, **[1]**7–8,
       [2.7]16–17
Biomechanical principles, **[1]**9–11
Breaks from work
  computer user, **[2.7]**23
  organizational controls, **[1]**36

C

Cardiovascular system, **[1]**13
Carpal tunnel syndrome, **[1]**18
  absentee rates, **[1]**27
  in office workers, 1–29
  symptoms, **[1]**22–23
Case study approach, **[1]**37

Cell physiology, **[1]**12–13
Certification of ergonomist, **[2.7]**20–21
Chemical industry
  anthropometrics in, **[2.2]**2–4
  hand injury in quality control laboratory,
       [2.2]6–8
Clerical work, see Office environment
Commerical facilities, mini-mall door
       opening, **[2.6]**2–7
Computer workstations
  evaluation of components, **[2.7]**23–24
  video display terminals, see Video display
       terminals
Congenital defects as predisposing factor in
       occupational disorders, **[1]**26
Construction, **[1]**5, 29–31
Control and mitigation strategies, **[1]**33–36
  environmental control methods,
       [1]34–35
  equipment--facility control methods,
       [1]35–36
  human control methods, **[1]**33–34
  organizational control methods, 36
Costs of workplace injury and illness, **[1]**5–8
  benefits of ergonomics programs,
       [1]7–8
  legal, **[1]**6
  medical, **[1]**5–6
  to organizations, **[1]**6–7
  of repetitive stress injury claims, **[2.5]**21,
       [2.7]17
Cubital tunnel syndrome, **[1]**18
Culture of workplace, **[1]**10
Cumulative trauma disorders, **[1]**17–19
  causes of, **[1]**26–27
  symptoms, **[1]**22–23

D

Definition of ergonomics, **[1]**1–2
Delivery services
  frozen food
    depalletizing manual materials handling,
       [2.8]2–7
    freezer loading, **[2.8]**9–15
    mini-mall door opening, **[2.6]**2–7
    parcel delivery facility, **[2.8]**18–20

Depalletizing, [2.8]2–7
DeQuervain's disease, [1]18
Design
    engineering, see Engineering design
    job design in hog industry, [2.1]2–6
    product design industry, [2.5]38–40
    redesign of facility for impaired worker,
        [2.8]]28–32
    tool design hazards, [2.8]22–26
    Vermont DMV case study, [2.7]25–34
Digital neuritis, [1]18
Disk herniation, [1]17, 20
Door of mini-mall, [2.6]2–7

E

Elbow injury, [1]18
Electronic monitoring of clerical work,
        [2.4]13–17
Electronics industry
    material handling issues, [2.3]2–4
    repetitive motion problems in manufactur-
        ing facility, [2.3]6–8
Engineering design
    benefits, [2.7]16–17
    certification of ergonomist, [2.7]20–21
    costs of repetitive trauma disorder,
        2.7]17
    epidemiology, [2.7]17–18
    issues in ergonomics, [2.7]22–24
    practice of ergonomics, [2.7]21
    social-legal-regulatory aspects, [2.7]19–20
    Vermont DMV case study, [2.7]25–34
Entry door study, [2.6]2–7
Environmental control methods, [1]34–35
Environmental monitoring of stressors, [1]11
Epicondylitis, [1]18
Epidemiology
    repetitive trauma disorder, [2.7]17–18
    work-related illness, [1]3–5
Equipment--facility control methods,
        [1]35–36
Ergonomic hazards, defined, [1]2
Ergonomics
    areas of importance, [1]8–10
    case study approach, [1]37
    causes of cumulative trauma disorders,
        [1]26–27
    control and mitigation strategies,
        1]33–36
        environmental control methods,
            [1]34–35
        equipment--facility control methods,
            [1]35–36

human control methods, [1]33–34
        organizational control methods, 36
    cost of workplace industry and illness,
        [1]5–8
    definition of, [1]1–2
    existing solutions, [1]36–37
    high-risk workplaces, [1]27–33
        construction, [1]29–31
        manufacturing and assembly,
            [1]29–33
        meatpacking, [1]27–28
        office environment, [1]28–29
    importance of, [1]2–4
    incidence of workplace injury and illness,
        [1]4–5
    issues in, [2.7]22–24
    physiological factors, [1]11–16
        cardiovascular, [1]13
        cellular, [1]12–13
        multiple system involvement, [1]16
        musculoskeletal system, [1]15–16
        nervous system, [1]13–15
    practice of, [2.7]21
    problem identification strategies,
        [1]10–11
    types of injuries, [1]17–27
        cumulative trauma disorders,
            [1]17–19
        musculoskeletal disorders,
            [1]19–25
        physiological stress disorders, [1]25
Ergonomist
    certification of, [2.7]20–21
    defined, [1]2
Existing solutions, [1]36–37

F

Food service industry
    frozen food delivery
        depalletizing, [2.8]2–7
        freezer loading, [2.8]9–15
    restaurant chain
        redesign of facility and equipment for
            impaired worker, [2.8]28–32
        tool design hazards, [2.8]22–26
Forearm entrapment syndrome, [1]18
Frequency of work-related illness, [1]2

G

Ganglionic cysts, [1]19, 23
Group behavior, [1]10
Guyan tunnel syndrome, [1]18

H

Hand injury, **[1]**18
Handpacking system, **[2.5]**30–33
Health care industry
    back injury
        in hospital, **[2.4]**2–4
        manual lifting case studies, **[2.4]**10–11
        psychosocial case study of transcription
            unit, **[2.4]**13–17
        signal detection in, **[2.4]**6–8
Herniated disk, **[1]**20
Herniated disks, **[1]**17
High-risk workplaces, **[1]**27–33
    construction, **[1]**29–31
    manufacturing and assembly, **[1]**29–33
    meatpacking, **[1]**27–28
    office environment, **[1]**28–29
Hog industry
    job design, **[2.1]**2–6
    overexertion in, **[2.1]**8–11
Hospital
    back injury elimination in, **[2.4]**2–4
    signal detection in, **[2.4]**6–8
Human control methods, **[1]**33–34
Humidity, **[1]**11

I

Incidence of workplace injury and illness,
    [1]4–5

J

Job activity analysis, **[1]**10
Job design in hog industry, **[2.1]**2–6
Job rotation, **[1]**36

L

Laboratory workers, chemical plant quality
    control, **[2.2]**6–8
Legal issues
    costs of work-related injury and illness,
        [1]3, 6
    repetitive trauma disorder, **[2.7]**19–20
Lifting, see also Materials handling
    injuries caused by, **[1]**19–22
Lifting Equation, **[1]**20–22, **[2.8]**12
Lifting index, **[2.8]**12
Ligament sprains, **[1]**17
Lighting, **[1]**11
    for computer use, **[2.7]**23
    equipment control methods, **[1]**35
Livestock, see Hog industry

M

Manufacturing and assembly, **[1]**5, 29–33
    automobile radiator parts facility,
        [2.5]17–19
    back injury, manual lifting case studies,
        [2.4]10–11
    behavioral engineering approach in,
        [2.5]2–12
    electic motor, **[2.3]**6–8
    electronics plant, **[2.3]**2–4
    handpacking systems, **[2.5]**30–33
    product design industry, **[2.5]**38–40
    servo-motor assembly facility,
        [2.5]35–36
    synthetic fibers plant, **[2.5]**21–27
Materials handling
    construction trades, **[1]**30
    costs of, **[1]**7
    in electronics industry, **[2.3]**2–4
    food delivery service
        depalletizing, **[2.8]**2–7
        freezer loading, **[2.8]**9–15
    injuries caused by, **[1]**19–22
    manufaturing and assembly, **[1]**31
    meatpacking industry, **[1]**27, 28
Meatpacking, **[1]**5, **[1]**27–28
Mechanical back syndrome, **[1]**17
Medical costs of work-related injury and
    illness, **[1]**3, 5–6
Medical status as predisposing factor in
    occupational disorders, **[1]**26
Mercantile case studies, mini-mall door
    opening, **[2.6]**2–7
Mini-mall door opening study, **[2.6]**2–7
Monitoring, clerical workers, **[2.4]**13–17
Motivation, **[1]**10
Muscle strain, **[1]**17, see also Materials
    handling
Musculoskeletal system, **[1]**15–16
    disorders of, **[1]**19–25
    predisposing factors in occupational
        disorders, **[1]**26
Myalgia, **[1]**18
Myofascial syndrome, **[1]**18

N

Nervous system, **[1]**13–15
Neuritis, digital, **[1]**18
NIOSH Lifting Equation, **[1]**20–22,
    [2.8]12
Noise, **[1]**11, 34
Nuclear facility accident, **[2.9]**2–5

O

Occupational overuse syndrome, see Cumulative trauma disorders; Repetitive motion injury
Office environment, [1]5, 28–29
  ergonomics and engineering design, [2.7]16–34
    benefits, [2.7]16–17
    certification of ergonomist, [2.7]20–21
    costs of repetitive trauma disorder, [2.7]17
    epidemiology, [2.7]17–18
    issues in ergonomics, [2.7]22–24
    practice of ergonomics, [2.7]21
    social-legal-regulatory aspects, [2.7]19–20
    Vermont DMV case study, [2.7]25–34
  psychosocial case study of medical transcription unit, [2.4]13–17
  video display terminals
    in agricultural industry, [2.7]11–14
    ergonomic complaints, [2.7]2–6
    solution of problems, [2.7]8–9
Organizational control methods, 36
Organizations, costs of work-related injury and illness, [1]5–7
OSHA fines, [2.7]19–20
Overexertion, [2.1]8–11

P

Pallets, depalletizing, [2.8]2–7
Parcel delivery facility, [2.8]18–20
Performance monitoring, electronic, [2.4]13–17
Personal protective equipment, [1]10, 34, [1]35
  behavioral engineering approach, [2.5]2–12
  organizational requirements, [1]36
Physiological factors, [1]11–16
  cardiovascular, [1]13
  cellular, [1]12–13
  multiple system involvement, [1]16
  musculoskeletal system, [1]15–16
  nervous system, [1]13–15
Physiological stressors, [1]9–11, [1]25
Power plants, Three Mile Island nuclear facility accident, [2.9]2–5
Problem identification strategies, [1]10–11
Product design industry, [2.5]38–40
Productivity killers, [1]27

Productivity losses, [1]7
Productivity monitoring, electronic, [2.4]13–17
Product quality gains, [1]8
Pronator teres syndrome, [1]18
Prototype development and assembly, [2.5]38–40
Psychological factors, [1]9
Psychosocial factors, [1]10
  case study of medical transcription unit, [2.4]13–17
  resistance to change, [2.7]25
Push-pull-lift injuries, see Materials handling

R

Radial tunnel syndrome, [1]18
Raynaud's syndrome, [1]18–19
Recommended weight limit (RWL), [1]20–22
Regulatory issues, repetitive trauma disorder, [2.7]19–20
Repetitive motion injury, see also Cumulative trauma disorders
  chemical plant quality control laboratory, [2.2]6–8
  construction trades, 1–29, [1]30–31
  electric motor manufacturing facility, [2.3]6–8
  engineering design considerations, [2.7]16–34
    benefits of ergonomics, [2.7]16–17
    costs of, [2.7]17
    evaluation of workstation, [2.7]23–24
  handpacking systems, [2.5]30–33
  manufacturing and assembly, [1]31, 32
  meatpacking industry, [1]27
  office environment, [1]28–29
  synthetic fibers plant, [2.5]21–27
    employee training and involvement, [2.5]22
    exercise, [2.5]24
    furniture, equipment, and tool changes, [2.5]23
    program, [2.5]21–22
    psychosocial factors, [2.5]26
    strategies to eliminate, [2.5]25–26
    worksite analysis, [2.5]22–23
Restaurant chain
  redesign of facility and equipment for impaired worker, [2.8]28–32
  tool design hazards, [2.8]22–26
Rest breaks, [1]36
Risk factors, defined, [1]2

Rotation, job, [1]36
Rotator cuff tendonitis, [1]18

S

Segmental vibration, [1]23–24, 35
Service sectors
  frozen food delivery
    depalletizing manual materials handling,
    [2.8]2–7
    freezer loading, [2.8]9–15
  parcel delivery facility, [2.8]18–20
  restaurant chain
    redesign of facility and equipment for
    impaired worker, [2.8]28–32
    tool design hazards, [2.8]22–26
Servo-motor assembly facility, [2.5]35–36
Shoulder injury, cumulative, [1]17–18
Signal detection
  in hospital, [2.4]6–8
  Three Mile Island nuclear facility accident,
  [2.9]2–5
Social issues, repetitive trauma disorder,
  [2.7]19–20
Sprains, [1]17, [1]29, see also Materials
  handling
Statistics, occupational safety and health,
  [1]2–5
Strain injury, [1]17, see also Materials handling
  construction trades, [1]29
Stressors, [1]2
  environmental monitoring of, [1]11
  noise, [1]11, 34
  physiological, [1]9–11, 25
  temperature, [1]11, 34
Synergistic factors, [1]1–26
Synthetic fibers plant, [2.5]21–27
Systemic conditions as predisposing factor in
  occupational disorders, [1]26

T

Task analysis, [1]10
Temperature stress, [1]11, 34
Tendonitis
  absentee rates, [1]27
  rotator cuff, [1]18
Tennis elbow, [1]18
Tenosynovitis, [1]18
Terminology, [1]2
Thoracic outlet syndrome, [1]17–18, 23

Three Mile Island nuclear facility accident,
  [2.9]2–5
Time and motion studies, [1]10–11
Tool design hazards, [2.8]22–26
Trigger finger, [1]18
Tunnel syndromes, [1]18
Types of injuries, [1]17–27
  cumulative trauma disorders, [1]17–19
  musculoskeletal disorders, [1]19–25
  physiological stress disorders, [1]25

U

Unsheathed tendons, [1]18
Utility industry, Three Mile Island nuclear
  facility accident, [2.9]2–5

V

Vermont DMV case study, [2.7]25–34
  alliances, [2.7]26
  initial architectural design, [2.7]27–28,
    30–31
  involvement in design, [2.7]26
  solutions, [2.7]28, 29, 32–33
  trade study, [2.7]27
Vibration injury
  construction trades, 1–29
  meatpacking industry, [1]27
  segmental vibration, [1]1–24
  types of vibration, [1]24–25
  white finger (Raynaud's syndrome),
    [1]18–19, 23
Video display terminals, [1]28–29
  evaluation of components, [2.7]23, 24
  in agricultural industry, [2.7]11–14
  ergonomic complaints, [2.7]2–6
  solution of problems, [2.7]8
Videotaping, [1]10

W

White finger, [1]18–19, 23
Whole-body vibration, [1]23–25
Whole-job approach, [2.5]4–5
Work cycle analysis, [1]10–11
Workplace design, hog industry, [2.1]2–6
Work rate, [1]36
Work task analysis, [1]10
Wrist injury, [1]18

0 9 8 7 2 5 4 4 0

8 9 n r t 2 5 r r 8 0